From Television to
Home Computer

From Television to Home Computer

The Future of Consumer Electronics

Edited by

Angus Robertson

BLANDFORD PRESS
Poole   **Dorset**

First published in the UK in 1979

Copyright © 1979 Blandford Press Ltd,
Link House, West Street,
Poole, Dorset, BH15 1LL

ISBN 0 7137 0973 1

British Library Cataloguing in Publication Data

From television to home computer.
   1. Household appliances, Electric
   2. Electronic apparatus and appliances
   I. Robertson, Angus
   621.381   TK7018

Phototypeset in 10/12 Baskerville by Oliver Burridge & Co. Ltd,
Crawley, Sussex
Printed in Great Britain by Butler & Tanner Ltd,
Frome, Somerset

Contents

vi

Foreword

Twenty years ago, the only items of complex electronic equipment in a typical household would have been a monochrome television receiver and simple record player (at that time called a gramophone). Ten years later, colour television was just beginning to replace monochrome, and hi-fi 'separates' were becoming all the rage. Today the range of electronic equipment being introduced into the consumer environment is rapidly increasing and effectively opening up a vast new industry – that of 'Consumer Electronics'. Almost two thirds of all households now have a colour television, and a variety of 'peripherals' are being introduced that vastly increase the television's use from that of simply watching broadcast programmes.

These extras include video cassette recorders, which allow television programmes to be recorded for later viewing or permit pre-recorded cassettes (such as feature movies) to be watched. In addition, there are teletext and viewdata receivers (usually built into normal television sets) that provide access to vast data bases of consumer and business information such as news, weather, sport and entertainment guides. There are video discs that permit feature movies to be purchased for less than £10 and which are viewed by means of a record player that plays back pictures as well as sound (probably available by 1981).

Modern television and electronic games are allowing rapidly advancing electronic technology to provide entertainment of many levels, for adults and children, and basically, home computers are not unlike the more complex television games. However they have an important difference in that they may be programmed to perform many thousands of different tasks – in addition to providing games, of course.

Television receivers are themselves changing, with new facilities such as remote control and electronic tuning becoming available as standard fixtures on many sets. Also, televisions are now available with vast screens, using projection television techniques; and we have looked closely at aerials since without an aerial, it is rarely possible to satisfactorily receive television pictures.

Many other aspects of consumer electronics are also covered in this book, including electronic watches and calculators. In this field, a total revolution has taken place in only the last few years, so much so that many would find life tiresome without their trusty pocket calculator. In addition, with so much electronic equipment installed in the house, home protection and security is becoming increasingly important, not only from the point of view of theft, but also from fire which can be even more unpleasant.

Amateur radio has been established for about 50 years, but it is still used primarily by enthusiasts. However, citizens band (CB) radio (which will probably be introduced in Britain in the next few years) allows anyone, upon payment of a small licence fee, to participate in the hobby.

Finally, developments are continuing all the time in hi-fi, and these are thoroughly covered together with the increasingly important in-car entertainment market.

Consumer electronics are intent on changing and improving our lives over the coming years, and *From Television to Home Computer* provides an insight into that future.

A.R.

List of Contributors

DONALD ALDOUS   Technical editor of *Hi-Fi News and Record Review* and a specialist writer in audio technology.

JOHN ATKINSON   Deputy editor of *Hi-Fi News and Record Review* and a specialist writer in audio technology.

ROGER BUNNEY   Specialist writer in long distance television and aerial systems.

CHRISTOPHER CHAMBERS   Consultant specialising in security systems.

RICHARD ELEN   Editor of *Sound International* magazine, and a specialist writer in communications and electronic technology.

ADRIAN HOPE   Freelance writer in audio and video technology.

BARRY HUDSON   Managing Director of *Video Inclusive*, and a specialist in home video production.

ANGUS ROBERTSON   Editor of *Studio Sound* magazine and *Video Yearbook*, and a specialist writer in video and electronic technology.

CHRIS WEBB   Overseas correspondent on *Electronics Weekly* magazine.

Acknowledgements

The Editor and Publisher would like to thank the following organisations and individuals for use of material and illustrations:

Acoustical Manufacturing, Acoustic Research, Adam Imports, Advent Corp, Ampex, Apple, Atari, Barco, Bell & Howell/Avicom, BBC Ceefax, BBC EID, Blaupunkt, Brio Scanditoy, British Post Office, Casio, CBM, Cherry Leisure, CW Cameron, Decca, EMI Videograms, Eumig, GEC Telecommunications, General Instruments, Granada TV Rentals, Grundig, Hewlett-Packard, Hitachi Denshi, IBA, Ingersoll, Interton, Intervision, ITT Consumer Products, Jaybeam, JVC, KEF Loudspeakers, Labgear, Link House Studios, London Weekend Oracle, Milton Bradley, Mitsubishi, Mullard Research Labs, National Panasonic, Optimisation, Philips, Premier, Radio Rentals, Radofin, Rank Audio Visual, Revox, REW, Sharp, SME, Sony, Spectrum Marketing, Spendor, Speywood Communications, TDK, Technics, Teleng, Telefunken, Television International, Texas, Thorn Consumer Electronics, Unik Time, VCL, Video Inclusive, Videomaster.

1 Consumer Electronics – an Introduction

Angus Robertson

Why consumer electronics? Each of the two words has a specific meaning – consumers are those of us who buy household and personal goods manufactured by others, while electronics is a form of technology only understood by boffins. As a result you might say when those boffins become bored with designing computers and such technological marvels, they could devote their energies to designing electronics specifically for a much larger market – that of consumers, and so consumer electronics becomes reality.

Although consumer electronics is a readily accepted expression in the USA, as yet the term is relatively new to Britain; we spent many hours examining other possibilities for a suitable title for this book. 'Home Entertainment' or 'Electronic Entertainment' were possibilities but they provide a restriction to the home, and in-car entertainment is becoming a substantial market. Entertainment does not really 'fit' either, since most products offer many other purposes than straight entertainment value (such as education and culture), while electronic calculators provide labour saving, and electronic security systems also contribute towards safety. There are other areas where consumer electronics is rapidly making an impact, such as in the 'white goods' area of consumer appliances. Appliances have used electrical timers and actuators for years, but now this rudimentary electrical circuitry is being replaced by electronics which enables considerably more complex functions to be performed with improved reliability. Typical applications are in washing machines and dish washers, where a complex sequence of events requires control, and in central heating timers where electronic time clocks can operate the system during differing times each day, all pre-programmed.

Before proceeding to examine the numerous areas of consumer

electronics covered by this book, perhaps it would be interesting to investigate the economics behind this billion pound business. Fig. 1.1 shows that in a breakdown of total consumer expenditure, 27·9% of income is termed 'discretionary expenditure' for what one tends to

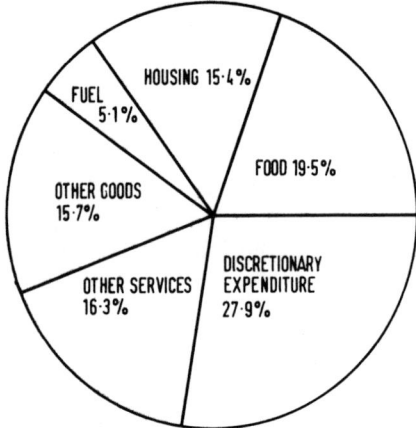

Fig. 1.1 Total consumer expenditure shown as percentage of income

think of as 'luxuries'. A further breakdown of discretionary expenditure is indicated in Fig. 1.2 which shows that clothing, alcohol and tobacco account for over 70%, leaving only 8·2% for consumer electronics. Of course, economic changes bear most heavily on discretionary spending, most of which carries heavy excise duties which are almost invariably raised each year. Nevertheless, the cost of typical consumer electronics products tends to remain remarkably stable

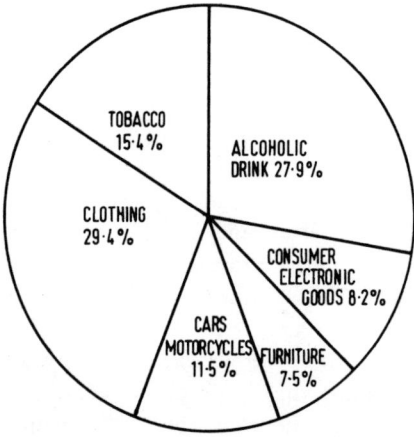

Fig. 1.2 Breakdown of discretionary expenditure which is usually subject to economic changes such as tax increases

while inflation pushes the cost of most other expenditure upwards. Fig. 1.3 shows how colour television prices have risen only 20% since 1970 yet include many new features, while the cost of motor vehicles has risen 150% and housing 130%. Of course, there have been examples of consumer electronic goods, such as pocket calculators, that have in fact fallen from £80 to less than £5 in the same period – but this is unusual and occurred primarily due to mass production, small mass per unit and lack of moving parts (since calculators are pure electronics mounted in a plastic case). Thus, although development cost of the integrated circuits used in calculators is very high, unit cost is very small.

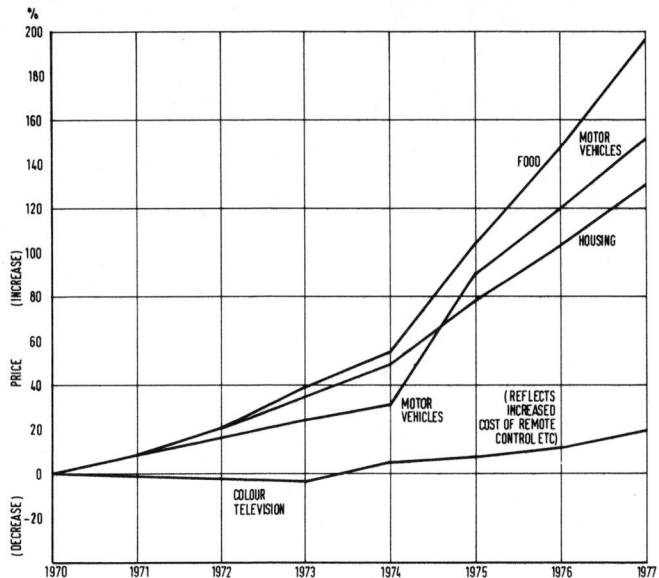

Fig. 1.3   Analysis of price increases since 1970 of food, motor vehicles and housing compared to colour television

Fig. 1.4 indicates how purely 'consumer' spending power has increased very considerably since 1970, so much so that a colour television purchase would have required 10·25 weeks average wages in 1970, being reduced to 3·6 weeks in 1978. This has brought colour television and many of the other more advanced consumer electronics products within the reach of many more families.

The total value of the consumer electronics market was estimated at

3

Fig. 1.4 Consumer spending power

	Average weekly wage[1]	Colour TV cost[2] (£)	No. of weeks wages per colour TV
1970	28·05	287·60	10·25 weeks
1971	30·93	289	9·34 weeks
1972	35·82	272·60	7·61 weeks
1973	40·92	264	6·45 weeks
1974	48·62	264	5·06 weeks
1975	59·68	274·10	4·60 weeks
1976	66·97	281·94	4·21 weeks
1977	72·89	285	3·91 weeks
1978	83·09	299	3·6 weeks

[1] Average wage for men, 21 years and over.
[2] Early TV prices were recommended prices; later prices are typical in shops.

just over £1bn in 1978, rising to over £2bn over the following seven years. For comparison, the motor vehicle market was worth about £3bn in 1978, of which a considerably proportion is business operated; so it can be seen that consumer electronics is a very substantial market. Fig. 1.5 gives a complete breakdown on estimated market values for the various consumer electronics product lines and from which a number of very intriguing trends can be observed. The television receiver market remains almost stationary, while radio receivers and in-car entertainment show gradual increases until 1985. On the other hand, the market for hi-fi separates (record turntables, tape recorders, amplifiers and tuners, and loudspeakers) gradually falls, but this is made up by sales of music centres which show an increase of about 80% over the next seven years. However, the three markets where phenomenal expansion will be seen are video cassette recorders (from £10m in 1976 to £400m in 1985), teletext and viewdata receivers

Fig. 1.5 Estimated market values for consumer electronic products

Product	1976	1978	1980	1985
Television receivers	£450m	£466m	£443m	£454m
Radio receivers	£87m	£110m	£123m	£147m
In-car entertainment	£47m	£50m	£57m	£60m
Video cassette recorders	£10m	£52m	£165m	£400m
Record turntables	£75m	£73m	£64m	£62m
Tape recorders	£78m	£56m	£53m	£50m
Amplifiers and tuners	£28m	£27m	£27m	£25m
Loudspeakers	£36m	£38m	£34m	£30m
Music centres	£132m	£177m	£227m	£322m
Teletext and viewdata	£0·1m	£12m	£100m	£400m
Television games	£1·4m	£15m	£30m	£200m
Totals (rounded)	£950m	£1·08bn	£1·33bn	£2·16bn

(£0·1m in 1976 to £400m in 1985) and finally television games (£1·4m in 1976 to £200m in 1985). One possible market expansion area not indicated in this table is that for video disc players. Unfortunately, there are far too many variables to predict accurately a market size (for instance until a substantial number of videograms or video disc programmes are available, very few video disc players will be sold, and of course vice versa). After many years of stop-go announcements, the video disc player has finally arrived on the USA market with a price tag of $695 and first run feature movies costing only $15·95 to *buy*. With an average cinema admission price in the USA of $4, one is now able to take the movie home for repeated viewing, for the same cost as taking the average family just once, and that does not include travelling. MCA has just completed construction of a video disc replication plant (pressing factory) for the manufacture of discs, but an economical replication run is a minimum of 5,000 and so until the European market can support this volume, successful marketing will present numerous difficulties.

All these new technologies require a colour television as a fundamental display device, so Fig. 1.6 shows the impact of television in Western Europe and the USA. The poor impact in certain countries can usually be traced to low gross national product, abetted by high import duties in these same countries (36% or 51% rather than the typical 14% of the EEC). What is perhaps surprising from Fig. 1.6 is that although Britain has 105% television receiver per household penetration (in other words some have more than one set); Britain also has the lowest GNP growth rate in Western Europe, and its GNP per head of population is very poor when compared to immediate neighbours such as France and West Germany. This phenomenon is explained by the British television rental system which is almost unique, accounting for over 60% of colour television receivers in the market place.

One other interesting factor emerging from Fig. 1.6 is the high level of television receivers in the USA. Referring now to Fig. 1.7 it can be seen that colour television first showed an impact in 1956, and reached 68 million sets by 1977. But what is interesting is that the monochrome television market has also been increasing during this period, unlike that of Britain (Fig. 1.8) where monochrome deliveries have been consistently falling from a peak of over 15 million in 1970, to less than 7 million in 1978, while colour television impact has now reached over 11 million receivers. This figure does not include households with

5

Fig. 1.6 Comparison of West European and USA populations, economies and television receiver saturation levels

	Area km²	Population 1976	Households 1976	Gross national product 1975	GNP real growth rate 74 on 70	No. of colour TVs in 1975 (percentage of households)		No. of monochrome TVs in 1975 (percentage of households)		Percentage of TV receivers per household 1975
Austria	84k	7·5m	2·63m	£21bn	5·7%	600k	(23%)	1·4m	(53%)	76%
Belgium/Lux	33k	10·2m	3·57m	£36bn	5·1%	480k	(13%)	2m	(56%)	69%
Denmark	43k	5·1m	1·79m	£18bn	2·8%	400k	(22%)	1·2m	(67%)	89%
Eire	70k	3·2m	1·12m	£5bn	5%	100k	(9%)	500k	(46%)	55%
Finland	337k	4·7m	1·65m	£15bn	5·5%	250k	(15%)	1·35m	(82%)	97%
France	552k	52·9m	18·13m	£190bn	5·2%	3m	(16·5%)	13m	(72%)	88·5%
West Germany	249k	61·5m	22·4m	£237bn	3·7%	8·5m	(38%)	11m	(49%)	87%
Italy	301k	56·2m	17·45m	£96bn	3·9%	500k	(3%)	11·5m	(66%)	69%
Netherlands	41k	18·8m	4·42m	£45bn	4·2%	1·5m	(34%)	2·1m	(47·5%)	81·5%
Norway	324k	4m	1·4m	£16bn	4%	200k	(14%)	900k	(64%)	78%
Portugal	92k	9·4m	3·3m	£7bn	6%	no colour		750k	(23%)	23%
Spain	505k	36m	12·63m	£48bn	6·3%	150k	(2%)	6·5m	(51%)	53%
Sweden	450k	8·2m	2·87m	£36bn	3%	1·45m	(51%)	1·45m	(51%)	102%
Switzerland	41k	6·4m	2·25m	£31bn	3·5%	500k	(22%)	1·3m	(58%)	80%
UK	244k	56m	19·5m	£128bn	2·7%	9m	(46%)	11·5m	(59%)	105%
USA	9363k	215m	75·9m	£840bn	2·5%	56·5m	(74%)	65·7m	(87%)	161%

Some households computed as 2·85 persons; some households own more than one television, thus higher than 100% saturation levels.

6

Fig. 1.7   USA television total market

Year	Colour TV	Mono TV
1955	—	37·6m
1956	100k	42·7m
1960	500k	55·1m
1962	1m	60·2m
1964	3·1m	64·1m
1966	10m	64·8m
1968	20·9m	62·7m
1970	31·6m	61·6m
1972	45·4m	64·4m
1974	57m	64m
1976	63·2m	68·3m
1977	68m	70·2m

Fig. 1.8   British television penetration based on broadcast receiving licences

Year	Colour TV	Mono TV
1967	—	14·267m
1968	20k	15·068m
1969	99k	15·397m
1970	273k	15·609m
1971	610k	15·333m
1972	1·635m	15·024m
1973	3·332m	13·793m
1974	5·558m	11·766m
1975	7·580m	10·120m
1976	8·639m	9·149m
1977	9·958m	8·098m
1978	11·519m	6·772m

Total for 1978, 18,290,780

Figures rounded, and there are an additional estimated 1,000,000 evaders with non-licenced television receivers; some households have more than
one TV, but only require one licence.

more than one colour television (nor the estimated one million households still dodging the Post Office television detector vans) and in fact
represents about 65% of British households with colour television.
This is expected to rise to 83% by 1983 when 99% of households will
have television receivers.

The colour television penetration (total) and market size are shown
in Fig. 1.9, and it may be noted that there are far more sets coming
into service than showing on the penetration. This is accounted for by
the replacement market since the colour televisions sold in the early
1960s are becoming obsolete, and because new receivers offer many

Fig. 1.9   Colour TV penetration and market size

Year	Colour TV penetration	Colour TV market
1975	9m	1·6m
1976	10·5m	1·5m
1977	12m	1·6m
1978	13·5m	1·7m
1980	15·5m	1·8m
1983	16·8m	2m
1985	17·3m	2·1m

Later colour TV sales are replacement models, often with new facilities.

other facilities such as remote control, teletext and viewdata. Now that the main colour television market is becoming saturated, manufacturers are beginning to promote portable colour television in an attempt to launch two colour televisions (and potentially more) into each household. Referring back to the USA television figures in Fig. 1.7, the large monochrome market is explained by second television sets, mostly portable, and while this has not yet occurred in Britain, it is estimated that over 3·5 million portable colour televisions will be in use in Britain by 1983. Oddly enough, these will be owned both by higher income households looking for a second set and lower income households looking for a cheap colour set (about 15% less than standard model) – middle income groups will have to be satisfied with a single colour set.

New colour television receivers often offer remote control (estimates show 60-70% offering this facility within three years), while teletext and viewdata will also create a considerable impact. In Britain, teletext is a free information service broadcast by both BBC and ITV, while viewdata is transmitted along telephone lines by the Post Office (PO) (although the PO does not supply the information) and is charged on usage. Fig. 1.10 indicates estimated markets for teletext

Fig. 1.10   Total penetration of televisions with teletext/viewdata and TV games

Facility	1978	1980	1983	1985
Teletext only	25k	200k	2·5m	5m
Teletext/viewdata	2k	50k	750k	2m
Simple non- and semi-programmable games	500k	1m	2·5m	5m
Fully programmable games	10k	100k	500k	2m

only receivers, and receivers offering both teletext and viewdata. There will also be a small market for viewdata only receivers, but this will be primarily business, and is not shown here. By 1978, teletext was available to every household in the country (and had been for three years) and being a free service, is only held back by the cost of teletext receivers. The additional circuitry required within a television to provide teletext is pure electronics and will come down rapidly in cost, just as that of calculators did a few years ago. From the present £650, it should fall in the early eighties to only £50 on top of television set cost, making a total of £350 to £400. When this point is reached, teletext will be as common as remote control today. On the other hand, being a paid service, viewdata will be restricted by the actual information being offered – in many cases this will only become economic for the companies supplying information when the market reaches a certain size. So in early years, a substantial proportion of the viewdata market will be business which can justify the cost of information more easily, while once a mid-point has been passed in receiver sales, the market will expand rather more quickly, to reach an estimated two million by 1985.

Television games are becoming very familiar, or rather the very basic football and tennis type games are becoming widely available. These are the simple non-programmable games, while a generation of semi-programmable types has arisen during 1978 which provide a restricted range of cartridges that plug-in to the main game and estimates of these reach five million in the market by 1985. Fully programmable games on the other hand provide for a wide range of different games on 20 or more cartridges, and are rather more complex than semi-programmable types, and more expensive probably coming down to around £50 by 1985 (although currently at over £100). The market estimate is for two million by 1985 if costs come down to this level.

Video cassette recorders will be a half billion pound market by the late 1980s, and some sources are suggesting a final market saturation of half that of the colour television market – in other words over eight million. Unfortunately, unlike domestic records and audio cassettes, there are currently four different video cassette formats being marketed in Britain – VHS from JVC (and others), VCR-LP from Philips, Beta from Sony (and others) and SVR from Grundig. Although figures provided by each of the four format 'families' obviously contradict the market share indicated in Fig. 1.11 is believed to be

9

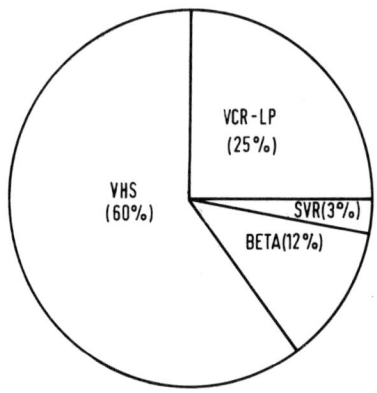

Fig. 1.11 Video cassette recorder estimated market shares in 1979 (based on imports and sales). The introduction in 1980 of the Philips V2000 format significantly affects these estimates

accurate. It is based on discussions with importers, retailers and the companies duplicating and supplying pre-recorded video cassettes. The VHS family (JVC, Akai, Hitachi, Panasonic, Thorn, etc.) come top with a 60% market share, a considerable proportion of which is due to rental, while VCR-LP from Philips with a four month lead come second with 25%. Beta from Sony, Sanyo and Toshiba come third with 12% (partially due to late marketing and lack of rental outlets), while Grundig SVR comes a poor fourth with only 3% despite its four hour recording time. Penetration estimates for video cassette recorders, based on research for which actual 1978 figures were made available, show 1978 as being a total of 100,000 video casette recorders, rising to 270,000 in 1979, 525,000 in 1980, 1,600,000 in 1983, and finally 2,700,000 by 1985 (Fig. 1.12). During 1978, 401,930 video cassette recorders were sold in the USA.

Fig. 1.12 Video cassette recorder estimated penetration and market sizes

	1978	1979	1980	1983	1985
Penetration	100k	270k	525k	1·6m	2·7m
Market size	75k	170k	255k	460k	665k

As mentioned earlier, Britain has always had a very high proportion of rented colour television receivers (Fig. 1.13). It can be seen that as the market becomes more confident of the reliability of colour television, and the price becomes within the reach of lower income households, that rented penetration falls from 69% in 1971 to an estimated 50% by 1984. This reduction in rental televisions has led the companies to become heavily involved with new areas in order to retain their profitability, and these include teletext and viewdata, and video cassette recorders. It is estimated that 65% of the video cassette recorders supplied next year will be on rental agreements. Currently, the three Thorn rental companies (Radio Rentals, DER and Multibroadcast), and Granada are all renting VHS format video cassette recorders, while Visionhire (partially owned by Philips) is perhaps, not surprisingly, renting Philips VCR-LP recorders.

Fig. 1.13   Rental market size for colour TV and video cassette recorders

	1971	1977	1979	1984
Colour TV	69%	61%	59%	50%
Video	—	—	65%	58%

Whereas colour television receivers are now virtually totally electronic, having finally removed the mechanical tuning controller and are thus inherently reliable, video cassette recorders remain based on complex mechanics (and electronics) and require regular servicing (which TV does not); and so renting is currently by far the favoured approach. Beta has as yet no national rental outlets, although a few independent dealers are known to be providing machines on rental, and Beta is unlikely to increase its market share in the long run unless this is organised. Since video cassette recorders are basically designed for recording television programmes off-air, it does not really matter if a variety of formats are in use, since there is little interchange of programmes between consumers. However, in the long term, the less popular recorders will not find economic video cassettes or videograms (video cassette programmes) available. Obviously, videogram suppliers would prefer to keep only one single format cassette in stock, since multiple formats means much larger stocks to cater for differing demand.

Having discussed these principle areas of market expansion, it is interesting to note from Fig. 1.14 the 'real' cost of purchasing these new consumer electronics products in terms of the cost of buying a colour television over the past eight years (which was fully detailed in Fig. 1.4). From Fig. 1.14, it can be seen that a video cassette recorder requires nine weeks average wages today (1978/1979), and that was equivalent to the cost of a colour television in 1971. In other words, if you bought a colour television in 1971, a video cassette recorder today will cost the same as the television did then in terms of wages.

Although teletext and viewdata, television games, and video cassette recorders are the main areas of interest in consumer electronics, there are numerous other areas which are also covered by this book. For instance, although the earlier figures referred to television games,

Fig. 1.14 Real price comparisons of recent products compared with the equivalent cost of a colour television during 1971–78

Product	1978 price	In stated years, weeks wages to buy colour TV
Standard colour TV	£299	requiring 3·6 weeks wages
Remote control colour TV	£350	requiring 4·2 weeks wages, CTV cost 1976
Colour TV with games	£420	requiring 5·1 weeks wages, CTV cost 1974
Standard+2nd small colour TV	£539	requiring 6·5 weeks wages, CTV cost 1973
Colour TV with teletext	£650	requiring 7·8 weeks wages, CTV cost 1972
Video cassette recorder	£750	requiring 9 weeks wages, CTV cost 1971

1978 average wage of £83·09 used for comparisons.

the market for 'electronic' games is also substantial. Basically, electronic games operate using the same technology as television games, but rather than using the television receiver as a display device, use internal lights or indicators, so releasing themselves from the household living room. Although some electronic games can only be described as toys, those such as chess are aimed directly at adults. In fact electronic chess is already a multi million pound market in Britain. Although the trusty 20in or 26in colour television receiver will be here until at least 1990, large screen television using television projectors is becoming significant. Although initially used for commercial applications where a television picture is to be presented to a large audience, television projectors are making an impact in consumer areas, and will also become a multi million pound market during the early 1980s. Television aerials are often overlooked, but are nevertheless extremely important, particularly when using teletext. Although cost is still high, home video production will also doubtless become an area of market expansion during the eighties.

Steering away from the television, home computers are currently aimed at a rather specialist market, but there are many applications in the home and small business for such small computers. Electronic watches and calculators has been a cut throat business causing many casualties along the way; but it is still booming. Home protection and security systems might not appear to be consumer electronics, nevertheless with the increasing petty crime and extremely high percentage of unsolved thefts, home security is as essential as ever, as are automatic fire alarms and such protective devices.

Citizens band (CB) radio has become over a $200m market in the USA, and although doubtless there is a substantial market in Britain, unfortunately the government in the form of the Home Office, will not yet permit CB to be introduced in Britain (1979) despite the numerous proven advantages. So the market remains non-existent. Although Britain, Belgium and Finland have all made CB illegal, there is much activity in other European countries, particularly France, West Germany, Sweden and Switzerland where CB is perfectly legal and for which a $50m market is forecast in 1980.

The final chapters provide coverage of sound, or hi-fi as many term it. There are several developments in broadcasting including experimentation with surround sound (previously known as quadraphonics) and Dolby-B noise reduction systems, while general developments in hi-fi are also examined. In-car entertainment is also becoming a substantial market, and indeed is virtually responsible for keeping the pre-recorded compact cassette alive and kicking. Stereo systems are becoming fashionable, with even the occasional four channel system. One other important area, not strictly entertainment at all, is the BBC Carfax motoring information service which should become available during the early 1980s.

So that is the consumer electronics market. In the following pages, I have written chapters about consumer electronics areas within which I specialise, while other professional journalists have kindly contributed chapters covering technology areas within which they each specialise. Special thanks to each of them, and also thanks to ITT for some of the consumer statistics used in this chapter*.

---

*Since the statistics for this book were compiled, value added tax (VAT) has been increased by the British Government. Thus it is likely that there will be a short term effect on sales of some goods in 1979, and some statistics may be slightly over-estimated.

2 Television Basics and Sound

Angus Robertson

Before looking in detail at the many aspects of television covered in later chapters, perhaps it would be best to examine the composition of the television signal since from this understanding, many techniques described later will become more obvious.

The whole basis of television picture transmission relies on two separate points – that there is a finite limit to the number of different images an eye needs to comprehend each second in order to follow continuous motion, and that provided the screen image remains relatively small and is viewed from a reasonable distance, this picture can be broken up into separate horizontal lines which visually merge with one another when seen on the television screen. Taking the first point first, even in the earliest days of films, it was known that about 50 separate pictures per second must be perceived for pictures to be seen without flicker, but filming at this rate would become very expensive on film stock. So a compromise is made and for the first silent movies a shooting rate of 16 frames per second was used, but this is artificially increased to 48 per second by using a triple shutter on the projector which interrupts the light source three times for each frame. In other words, each single film frame is flashed onto the screen three times. When sound movies came along a few years later, the shooting speed had to be increased to enable sound to be recorded with sufficient quality, and a 24 frame per second speed was arrived at with the projector shutters being modified to have only two blades – each frame was thus projected twice resulting in the same 48 picture per second rate. The same techniques is used for television pictures except that when the first television systems were developed, it was found convenient to lock the television frame rate to mains power frequency of 50Hz (or cycles per second as it was known in those days), so that 50 different pictures are transmitted each second. Locking the picture precisely to mains frequency also eliminated hum problems.

Fig. 2.1 shows how the television picture is divided into lines (in this case, for simplicity, only 11 lines). On the television display screen (cathode ray tube), a spot of light continuously traces out the pattern of lines being suppressed for flyback, in other words when it is returning to the left hand side of the screen ready to trace a further line. As previously explained, to eliminate flicker 50 pictures must be transmitted each second, but in order to reduce to bandwidth required for picture transmission (bandwidth being 'expensive'), the television picture is divided into two separate fields, each with exactly half the lines of the picture, in the case of Fig. 2.1, $5\frac{1}{2}$ lines in each field. So although we are transmitting 50 different fields each second, these make up only 25 different frames, making this the actual transmission rate. In a practical television picture, there are 625 lines, with $312\frac{1}{2}$ in each field although not quite all these lines are available for the television picture.

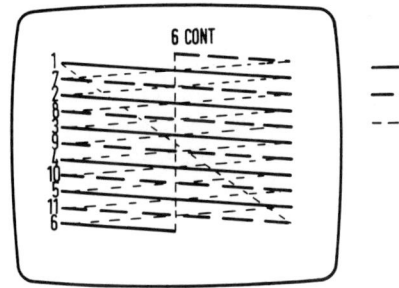

ODD SCAN
— — EVEN SCAN
— — — FLYBACK

Fig. 2.1   An 11-line interlaced TV picture

Since the television picture is composed of individual lines and fields, synchronising information is required to distinguish each line and field, and these are termed synchronising pulses and also allow time for flyback of the spot of light. Fig. 2.2 shows how the picture and synchronising information are combined to provide the television signal, which in this instance is a test waveform called sawtooth, i.e. a gradual transition from black through grey to white across the screen. Transitions are considered at about black level so that the line synchronising pulses fall below black, while information above black level is picture. At the end of each field, the spot of light has to return to the top of the screen in preparation for the next field, as can be seen in Fig. 2.1, so there is another synchronising period termed field syncs

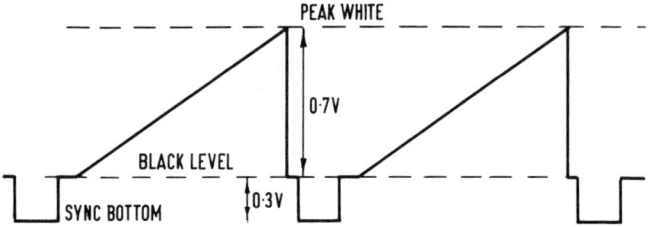

Fig. 2.2 Diagrammatic representation of two TV lines

which allows time for this to occur. Two spare lines are left at the end of one field, and 22 at the beginning of the next field which comprise this field sync period.

In practice, modern television receivers do not require all 24 such lines for field flyback, and so some of these spare lines are used for other purposes – lines 19, 20, 332, and 333 are used for insertion test signals which enable the performance of television signals to be automatically and continuously measured during normal transmission hours; lines 17, 18, 330, and 331 for the teletext information services described in Chapter 8; and line 16 for a BBC internal service called ICE (insertion communications equipment) and used amongst other things, for sending programme schedules from London to the regions. From the point of view of video recording, it is imperative that the synchronising information is recorded intact since otherwise the television receiver will be unable to lock-up to the pictures which will then shudder or in the worst case roll continuously.

As mentioned, the European television picture contains 625 lines transmitted 25 times per second. Each picture line contains some 400 picture points, each of which might be a different level of brightness (luminance) depending upon the scene content. Thus there are some 250,000 picture points in a television picture being repeated 25 times per second.

So the television signal comprises rapidly changing information requiring a considerable frequency bandwidth to be accommodated. In comparison, a high quality audio signal has a bandwidth of about 15kHz, but the television picture is some 350 times wider at 5·5MHz and is thus very much more difficult to transmit and record.

So far we have only considered monochrome television pictures since these fully illustrate television fundamentals, but next we shall consider colour. The earliest colour system comprised colour wheels

16

rotating before monochrome camera and television tubes, both in synchronism so that while the camera viewed the scene through a red filter, the television screen was also seen through a red filter. Although superseded many years ago by electronic techniques, the rotating disc was used for the colour cameras used by NASA in the Apollo moon missions because of simplicity. Yet in modern colour television cameras, instead of a single pick-up tube which is sensitive only to scene brightness, three separate tubes each with a filter, are used to separately distinguish red, green and blue colours (Fig. 2.3).

To transmit the three colours separately would be extremely wasteful of bandwidth (and impractical), so the colour signals are encoded and added to the basic luminance signal. There are three different systems of colour encoding used in the world, which will be detailed later; Britain and most of Europe uses the PAL system. To provide full compatibility for monochrome viewers, the luminance (brightness) signal is transmitted untouched and since this provides all the detail for the picture, the chrominance (colour) signal can be restricted in bandwidth, making it more easily accommodated within the luminance bandwidth. In practice, two colour difference signals (red minus luminance and blue minus luminance) from which red, green and blue signals can be decoded, are combined and modulated onto a suppressed carrier at 4·43MHz which is combined with the luminance signal (Fig. 2.4). Although extremely convenient for transmission, this technique does cause problems for consumer video cassette recorders because in these the bandwidth recordable is restricted so that the highest frequencies are lost, and the picture is not quite as sharp as it could be, although this is rarely noticed. Unfortunately, the chrominance signal is outside the recorded bandwidth. Special techniques, described in Chapter 3, are used to overcome this problem.

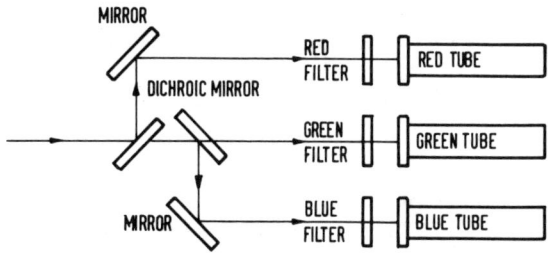

Fig. 2.3 Typical colour separation in colour TV cameras

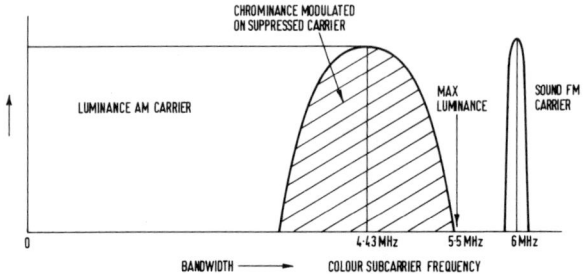

Fig. 2.4 Television signal spectrum showing liminance, chrominance and sound carriers

Television sound

Contrary to popular belief, television sound quality is almost as good as FM radio on Band 2. Unfortunately, the manufacturers of television receivers seem to have heard about some market survey (probably never conducted) that seems to say the British public are perfectly happy with low quality, tinny, distorted sound. Although various representations have been made to manufacturers, committees and broadcasting organisations, there are still virtually no televisions available in the UK that include a high quality sound output for a separate amplifier/speaker system. Thus it is necessary for those enthusiasts who care about the quality of their TV sound, actually to take the matter into their own hands. I shall describe later the various techniques for obtaining sound quality approaching that of the original broadcast, but first it is appropriate to consider the quality of sound transmitted by the broadcasting organisations.

Television sound is originated using similar techniques to those developed for radio. Broadcast quality microphones and audio mixers are used in TV studios, the sound quality not being limited or restricted in any way other than for purely artistic or operational reasons. In the case of ITV, sound is distributed to the transmitters using music quality PO audio lines while the BBC use a digital transmission system known as sound-in-syncs (S-in-S). S-in-S transmits the sound interleaved with the picture information to simplify distribution (no audio circuits required) and is similar to the pulse code modulation system used by BBC Radio for programme distribution. The digital sampling rate for S-in-S is 31,250 times per second which is only slightly lower than the PCM rate of 32,000 and to all intents and purposes quality is identical (bandwidth to 14kHz). The rather complex networking arrangements required for ITV programme distribution at present preclude S-in-S on purely economical grounds.

When the sound reaches the transmitter, it is handled separately from the video and transmitted on a frequency exactly 6MHz above the vision carrier. Vision is amplitude modulated while sound is frequency modulated with a deviation of 50kHz (slightly less than radio). The outputs of the vision and sound transmitters are combined immediately before the aerial, until which point they have been treated separately and the sound has been kept up to the same standards as radio.

It is true that certain compromises have to be made in the television studio, since microphones can not always be optimumly placed due to visual requirements, and background noise tends to be higher than for radio due to the level of activity in the studio itself and the powerful air-conditioning required to remove hundreds of kilowatts of heat radiated by the lights. Also, during editing of video tapes, audio tracks are copied from one tape to another (electronic editing) rather than being cut as in radio (mechanical editing). Audio performance of video tape recorders is not as high as audio only tape recorders as the recording tape is optimised for the video (which is considerably harder to record). Multitrack tape recorders are now often synchronised with VTRs resulting in improved audio performance, more versatile editing and stereo capability.

Incidentally, the sound bandwidth of all films transmitted using an opitcal sound track (principally feature films, serials and commercials) is limited to 7kHz by a historic specification for the track's pre-emphasis (or rather de-emphasis). It is interesting to note that when renting PO lines to link outside broadcasts and television centres together, only a bandwidth up to about 8kHz is commonly paid for, although it must be stressed that with careful equalisation it is usually much wider. Thus material originated locally (either studio or VTR) tends to have a higher quality than programmes originated in say Manchester and sent by PO lines to London for transmission (as with some ITV networked material). Since the BBC use S-in-S, quality is consistent throughout the UK (except for those homes receiving from repeaters or transposers).

Domestic TV receivers

So what happens to the sound when it reaches the domestic television set? Vision and sound were transmitted separately and ideally should be received separately. Unfortunately, it is considerably cheaper to follow the system shown in Fig. 2.5. The UHF tuner receives the Band 4 or 5 transmission (470MHz to 854MHz), amplifies and converts it to an intermediate frequency of 39·5MHz. Now since the sound carrier was transmitted 6MHz apart from the video carrier, it appears at an IF of 33·5MHz, just below that of the video (see insert (a) in Fig. 2.5). A single IF amplifier is now used to amplify both vision and sound to a level suitable for detection. Insert (b) in Fig. 2.5 shows the detected video with the sound carrier now at 6MHz. A second IF amplifier is tuned to 6MHz and the FM carrier then detected and demodulated down to audio. Finally, an audio amplifier drives a built-in speaker.

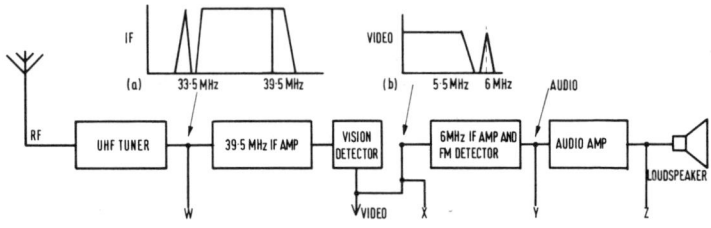

Fig. 2.5   Basic TV receiver and sound channel

However, there are a number of limitations to this design (used in virtually every TV set). The frequency response of the 39·5MHz IF amp is not usually particularly good at its extreme where the sound is located. A bandwidth of 3MHz is usually found adequate for monochrome sets although colour requires at least 5MHz. Since two separate signals are being amplified together, it is possible for the vision to interfere with the sound, particularly sharp edges of captions (which represent the highest frequencies in average picture content). This is known as intercarrier distortion and is inevitable with this design although its severity varies widely depending upon the accuracy of alignment of the IF stages.

The loudspeaker used in the TV set is limited both in size and quality since a heavy magnet would interfere with the scanning and colour registration of the picture. The plastic cases in particular that are sometimes used for cabinets tend to cause unpleasant colouration, and rarely is the drive unit properly acoustically loaded. Needless to say, the audio amplifier is rarely up to hi-fi standards, distortion of 5% is very common and double figures are not unusual.

High quality sound

There are four basic methods of obtaining improved audio performance of television sound:

1 Use an entirely separate tuner to pick-up the sound carrier. This could be a modified UHF tuner head which provides an IF frequency of around 100MHz suitable for receiving directly on a Band 2 tuner used as part of a standard hi-fi system (such a tuner head was described in *Hi-Fi News* April 1974). Motion Electronics manufacture a self-contained TV sound tuner with an audio output. Although theoretically providing the highest possible sound quality, these systems must be tuned separately to the television set and this is a severe disadvantage.

2 Use a separate IF amplifier for the sound signal (Fig. 2.6a). Input is taken from Point w which is the tuner head output. Sound would be detected directly at the IF amplifier output. As an alternative Fig. 2.6b uses a double conversion stage, converting the 33·5MHz carrier down to 10·7MHz from where it may be handled using standard FM radio modules and practices. Advantages of this system are that the IF strip may be tailored to have the correct bandwidth for the sound and that there is no intercarrier distortion.

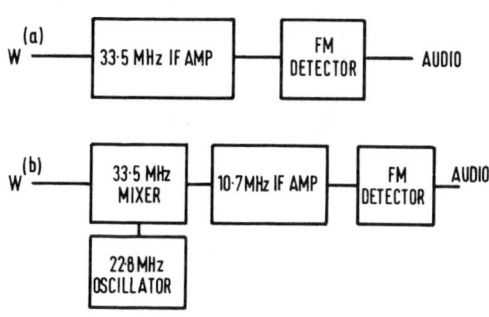

Fig. 2.6   Separate IF stages for TV sound channel

Using an IF transformer at the input isolates this sound IF from the remainder of the set and thus allows the audio output to be taken to a high quality amplifier. In Fig. 2.6b, as an alternative to mixing down to 10·7MHz, we could mix up to 100MHz and send straight into an FM tuner. Since sound is taken after the tuner stage, sound output will automatically follow the picture tuning. This technique is the most versatile and is used in professional TV tuners. However it is rather more complex than the system described under 4 below.

3 A coil may be used to pick-up stray radiation from the 6MHz IF amplifier which is then amplified and detected similarly to the TV set sound stage, Fig. 2.7. Since a pick-up coil is used, there are no problems of isolation or making connections to the TV set. In older sets, stray radiation made pick-up outside the cabinet possible but with recent types the pick-up coil often has to be physically placed within a couple of millimetres of the sound IF transformer or amplifier and even then only barely acceptable results may be found. The Celestion Telefi is an example of such a device and circuits using this technique have been published in various electronics magazines.

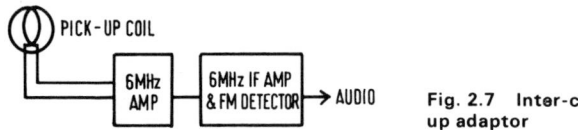

Fig. 2.7 Inter-carrier sound pick-up adaptor

4 Sound may be taken from the set after the detector but before it reaches the audio amplifier at Point y and fed into a separate high quality amplifier/speaker system. This overcomes the limitations of the set's audio section but is still prone to intercarrier distortion. However the simplicity and cheapness of the circuit described later warrants its use for many non-critical applications. Quality depends principally upon the alignment of the 39·5MHz IF amplifier and the response of the intercarrier IF amplifier and filters.

Isolation

That said, there is still one major problem. Another technique used by TV set manufacturers to produce the cheapest possible receivers is to eliminate the bulky, heavy and expensive mains isolating transformer used in audio equipment to make safe interconnection of equipment possible. Thus TV sets are connected directly to the mains and use either a dropper resistor to obtain the different voltages required, or more commonly in colour sets a switching power supply using a transistor or thyristor as an electronic switch. In all cases the TV chassis is above earth potential, either at neutral or even live. Thus, it is potentially dangerous to connect the chassis to audio equipment that may be earthed. The TV chassis must thus be isolated from the following audio equipment.

The most effective way is to use a mains isolating transformer and then earth the TV chassis. While this worked relatively well for monochrome sets which had power consumptions of 200VA or so, it was often difficult to mount the transformer inside the cabinet due to the magnetic field effecting CRT scanning. With colour sets the problem is compounded with purity (colour) problems. Manufacturers are quick to point out the low current consumptions of recent sets (100 to 120VA) but due to the switching power supplies used, the peak current consumption is usually between 500 and 1000VA and this is thus the size of transformer required (weighing around 9kg and having a large magnetic field).

As an alternative to such a large and bulky mains transformer, an audio isolating transformer could be used instead. Fig. 2.8a shows that this could be connected to either points Y or Z on Fig. 2.5 – if connected to point Z, a reasonably cheap loudspeaker isolating transformer may be used to connect an external high quality loudspeaker in place of the set's internal speaker. However, the drawback here is that the television sound has already been distorted by the internal power amplifier whose characteristics are generally tailored to match the internal amplifier. Most television rental companies will install such a transformer for about £4, one common application being to feed headphones or hearing aid for the hard of hearing. If an audio transformer is used to isolate the low level audio output from point Y, this must present a high bridging impedance to the set circuitry, otherwise this might be overloaded. Again, the transformer should be a high quality audio type with a bandwidth of 50Hz to 15kHz, and these cost from £5 up from companies such as Gardeners and Sowter.

Finally, possibly the cheapest technique for taking high quality sound from a television set is shown in Fig. 2.8b which uses an opto-isolator to provide a similar function as the transformer in Fig. 2.8a – but only at low level from point Y. Although an opto-isolator comes packaged in a small plastic case identical to an integrated circuit, it contains a light emitting diode driven by the audio, and a photo-transistor that detects the LED's varying intensity and converts this back into audio which may be fed to a separate power amplifier. Of course, light provides perfect electrical isolation enabling the external amplifier to be connected safely to the television receiver. Although this additional circuitry could be added to manufactured television receivers for less than £1, most manufacturers believe that the additional £3 or so that would add to the end cost of the receiver, to be totally uneconomic! Some European manufacturers produce sets with unusual power supplies that isolate the television chassis from the mains, and these sets typically have audio outputs built-in on a DIN connector. As yet, British manufacturers are not using this technique and the techniques already described must be used to remove sound safely.

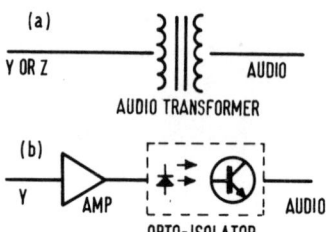

Fig. 2.8   Audio isolation circuits

With some TV sets, interference might be found on the sound taken from points Y or Z. For instance, due to intercarrier distortion, it is possible that a buzz will be heard on the sound. The channel tuning control should be adjusted to provide maximum sound quality with the sharpest picture (usually a compromise). With some sets, it is possible that line scan power (15,625kHz) or field scan (50Hz) will also be induced on the audio, but amplifier scratch and rumble filters, respectively, should reduce the effect (although harmonics of 50Hz can be troublesome). It is possible that other interference may occur with some sets, and the only real way to eliminate this is by using a separate IF amplifier for the radio.

Finally, one rather obvious point that is often neglected. Hi-fi loudspeakers almost invariably have large to enormous magnets. Now, colour television sets do not like magnets since they interfere with the shadoe mask and cause purity errors. Hence, it is essential that loudspeakers are kept at least 30cm from the TV set; the ear is surprisingly tolerant of the sound location.

3 Video Cassette Recorders

Angus Robertson

Perhaps the first question that should be asked when discussing video cassette recorders is why bother to record? After all, video recorders are still very expensive (and likely to stay that way), and one's money could surely be spent much more usefully on something like a home computer! Video cassette recorders appeal to several different groups of people.

Fanatics of particular programmes – whether sports or serials – will buy a video cassette recorder enabling them to archive such programming. However do bear in mind that this is still relatively expensive costing about £5 per recorded hour, although of course the tape may be reused hundreds of times if permanent recordings are not required. Many groups of people work unsocial hours which means they miss most evening programmes and have to make do with odd schools programmes and test card during the day. A video cassette recorder with a suitable electronic timer can record some of these programmes and thus timeshift them to a more suitable viewing time.

Alternatively, even those of us that work relatively normal hours find a video cassette recorder invaluable for recording programmes that would have otherwise been missed, or even enabling specific programmes to be watched at more convenient times – particularly useful at Christmas when many meal times will no longer have to be scheduled around the TV programme schedule. Of course, it enables programmes on conflicting channels to be watched since one channel can be recorded while simultaneously watching a second. Finally, libraries of video cassette programmes are beginning to become available so one can now rent or buy feature films and other entertainment and documentary type programming to watch at home.

Video cassette recorder philosophy

As their name suggests, video cassette recorders are designed for recording video, more commonly termed television. All the consumer types, which are the principle concern of this chapter, are designed specifically for recording television programme off-air (that is directly from a television aerial) and provide an output that may be connected using a simple coaxial cable extension lead (supplied with the recorder) to a normal domestic television receiver, either monochrome or colour. Again, all the recorders discussed record colour television pictures, but can be used just as effectively with monochrome television. To enable the video cassette recorder to record programmes directly off-air, it contains a television tuner similar to that contained in a normal television receiver, and which usually has eight channel preselectors. These may be individually tuned into any UHF television channels, again just like many television receivers.

The output from the video cassette recorder is modulating on to a spare television channel, usually between channels 35 and 45, using a micro television transmitter, and this may then be tuned into, using a spare preselector button on the television receiver. If the set has a button marked 'video' or VCR, this is the one to use. Otherwise it does not really matter. In most cases, the dealer or rental company that installs the video recorder will tune your television receiver into the video cassette recorder 'modulator', but if this ever needs resetting, the easiest technique is to set the video cassette recorder into play mode using a pre-recorded tape, and then tune the television preselector until the picture appears – this should be about half way through the travel of the preselector. If no recorded tapes are available (if the recorder is new), it should be possible to tune into a picture that will look effectively black on the television screen, rather than the usual random flashes of light (noise) that are seen when no transmitter is being received. Fig. 3.1 shows the simple connections that are used. The television aerial downlead is plugged into the video cassette recorder 'aerial or antenna in' socket, while a coaxial extension cable is used to connect the 'aerial out' or 'to TV set' socket to the aerial socket on the rear of the television receiver. The lead supplied will probably only be a few feet long and this might appear to urge you to place the recorder near to the television, and often actually on top of the set. This is very unwise since television sets become very warm, tend not to have load bearing tops and stray magnetic fields from the receiver could possibly cause partial erasure of the video cassettes. It is much better to buy or make up a much longer aerial lead that enables the recorder to be placed in a rather more convenient position in the room. Your dealer will advise on the correct plugs and sockets required, since these vary with different models of recorders.

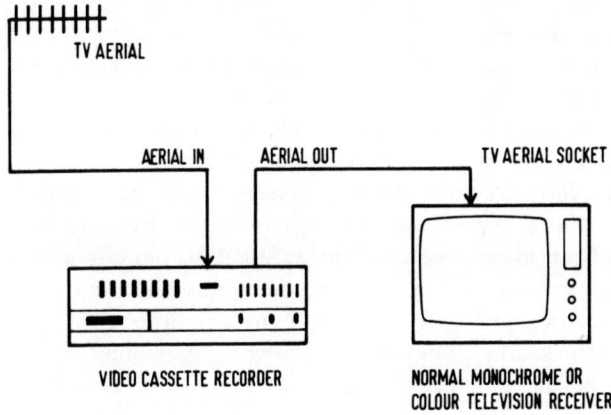

Fig. 3.1   Basic connections of video cassette recorder
and television

Consumer video cassette recorders may be used in three distinctly different modes:

1 Either manually or automatically (unattended using electronic timer) record a single television programme off-air;

2 Enable a single television programme to be recorded while the television receiver is used perfectly normally to watch the same or a different programme;

3 Enable a single television programme to be replayed and watched on the television receiver which should be tuned into the recorder's special channel.

One further application of video cassette recorders is that they may be used with a small television camera to record 'home movies'. Some recorders have special sockets enabling specialist accessories such as cameras to be directly connected, while others might require the use of an adaptor. In general, monochrome television cameras are low cost and operate with normal indoor light levels, while colour cameras invariably require substantial (over 2,000W) light indoors to provide satisfactory pictures, and cost from £1,000 upwards.

Before looking more closely at the various video cassette recorders

27

available, it is perhaps best to examine some of the techniques used for recording video and the difficulties encountered. Some readers might find this section slightly too technical, and could skip a few pages, following up at 'Early video cassette recorders' (page 34).

Operating principles

In essence, video cassette tape recorders operate much like normal audio tape recorders, but are probably umpteen times more complicated. The bandwidth of a normal audio channel ranges up to about 15kHz, while that for television pictures is about 5·5MHz, some 350 times greater. Normally, to enable high bandwidths to be recorded, the tape is run at a higher longitudinal speed and obviously such speeds are difficult to obtain – nevertheless, the BBC did develop an early video tape recorder called VERA in the fifties which ran tape at 200in/s, but it was superseded very quickly by a recorder developed by Ampex in the USA. Although now 25 years old, the mechanical transport format of the video tape recorders used by broadcasters today is identical.

The principle of video recording is simple. Since increasing the longitudinal tape speed is impossible, a ½in wide tape is used in consumer machines and the video recording heads made to rotate at high speed relative to the more slowly moving tape, thus laying a large number of long tracks at an angle across the tape (Fig. 3.2). Audio is recorded as a separate longitudinal track much as with a normal audio tape recorder, and a third control track is also recorded. This provides identification of the position of each video track so that when replayed, the video tape can be precisely positioned, otherwise the video heads might not cover the same path as those during the recording.

Early video tape recorders used open spools of video tape in much

AUDIO TRACK

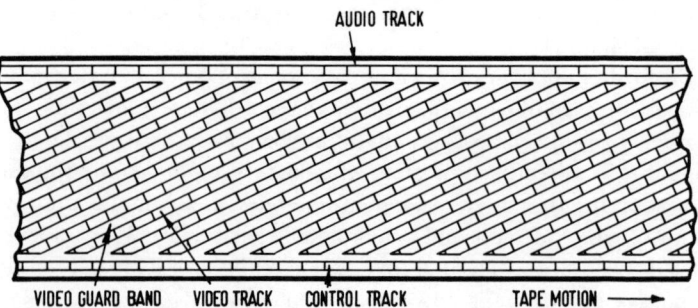

VIDEO GUARD BAND   VIDEO TRACK   CONTROL TRACK   TAPE MOTION ───▶

Fig. 3.2   Video tape track layout

the same way as tape recorders, but the complex rotating head assembly means that threading is rather complex, so much so that it is really impractical for consumer use. Hence the video cassette was developed. Although insertion of the cassette is simply accomplished, the video cassette recorder then has to remove tape from the cassette and thread it around the rotating video heads. Fig. 3.3 shows the unthreaded and threaded conditions of a typical video cassette recorder, in this case the Philips N1700. Basically, it can be seen that when the cassette is inserted, the video tape is located behind four guides mounted on a platter that can be rotated by a small motor until it reaches the position shown in the threaded diagram. It can be seen that the tape is then located around half of the head drum which houses the rotating video heads. In order that television pictures may be continuously recorded, there are actually two video heads on opposites of the drum which record alternate video tracks.

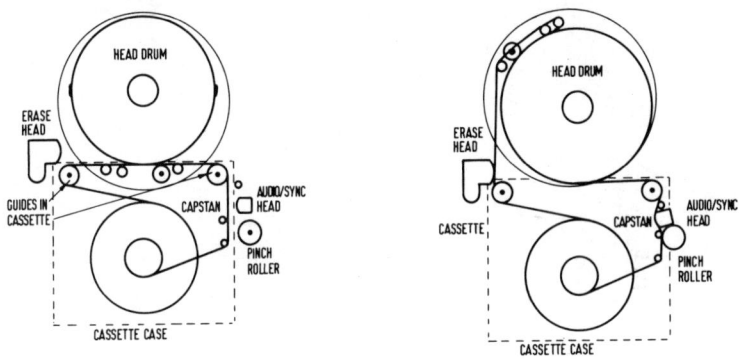

Fig. 3.3a   Philips N1700 unthreaded tape   Fig. 3.3b   Philips N1700 threaded tape

The audio and control tracks are recorded on a conventional looking audio head (but rather wider to cover both edges of the half inch tape) and a separate erase head covers the whole tape width. This type of tape threading where guides rotate is used on most video cassette recorders, but has the disadvantage of requiring a few seconds to thread. With this in mind, JVC developed a slightly faster loading system somewhat less mechanically complex using two parallel guides moving in a single direction (Fig. 3.4). Sony Beta format threading is similar to Philips, and the threaded tape path is shown in Fig. 3.5.

Because of these complex requirements for threading and rotating heads, video cassette recorders are extremely complex machines with

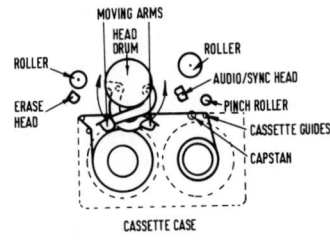

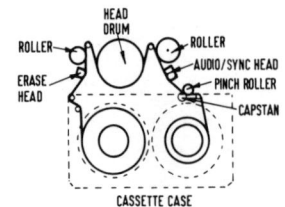

Fig. 3.4a   VHS unthreaded tape   **Fig. 3.4b   VHS threaded tape**

a vast number of expensively produced mechanical parts including at least three separate motors (tape drive, rotating head drum drive, and threading). This is the primary reason why the price of video cassette recorders is unlikely to be significantly reduced since mechanical

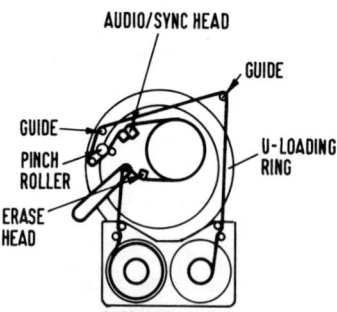

Fig. 3.5   Beta format threaded tape

components are heavily time consuming and will always tend to increase in price, unlike electronics which becomes cheaper due to miniaturisation, integrated circuits and more economical construction methods. Fig. 3.6 shows the interior of the JVC VHS – the head drum can be seen behind the cassette.

Electronics

The problems of economically recording colour television pictures took many years of research and it was only in 1973 that the first consumer oriented video cassette recorder was introduced by Philips. Late 1977 saw the introduction of the second generation of video cassette recorders which provided considerably improved recording time and these are discussed in this article.

Fig. 3.7 shows a schematic of the basic signal processing circuitry

30

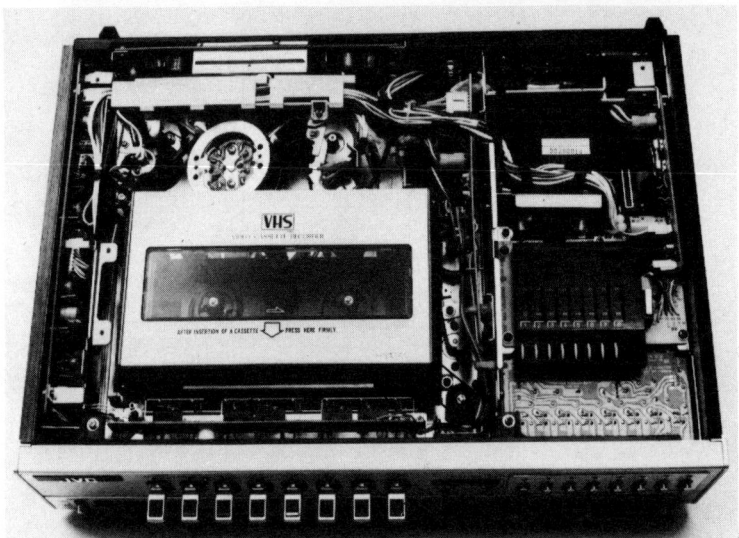

Fig. 3.6   Interior view of JVC VHS video cassette recorder showing head drum behind cassette holder, and rollers either side

blocks required for a consumer video cassette recorder with tuner. Since domestic television receivers have no facilities for video inputs or outputs, the video cassette recorder must be designed to operate connected into the aerial lead which is plugged into the recorder, and a lead taken from recorder to television set. Thus the recorder needs some form of aerial splitter, usually with a preamplifier, which feeds the television set and built-in TV tuner. The detector separates audio, luminance (the brightness of the picture) and chrominance (colour), which are then processed and recorded separately on the video tape using different techniques. Audio is recorded conventionally using a single head for recording and playback with an erase oscilator and head covering the full width of the half inch tape.

Recording television pictures is however somewhat more complicated. Fig. 3.8a shows the transmitted television spectrum which comprises an amplitude modulated 5·5MHz bandwidth luminance signal (the higher frequencies containing the finest detail of the picture) and the chrominance signal phase modulated onto a suppressed 4·43MHz carrier. Since all the picture detail is carried in the luminance, the chrominance bandwidth is restricted to about 1·5MHz. Although these high frequencies can be recorded directly onto video tape if the

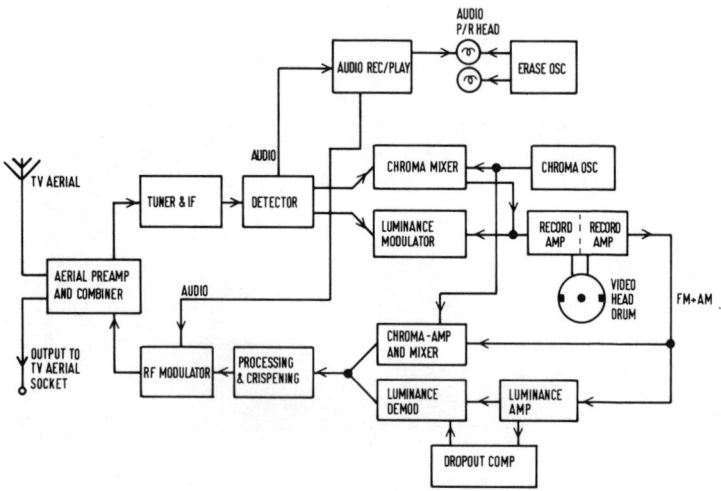

Fig. 3.7 Schematic arrangement of video processing for typical video cassette recorder

relative tape/head speed is sufficiently high, there is another limitation. Due to the wavelength of frequencies and the physical gap distance in any recording head, there is a theoretical limit of about nine octaves that may be recorded on magnetic tape. These nine octaves conveniently cover the audio spectrum between 50Hz and 20kHz, but is obviously insufficient to cover the 50Hz to the 5MHz of video frequencies. However, by frequency modulating this wide bandwidth video signal onto a carrier, all these frequencies can be effectively shifted up the spectrum where they can be very conveniently handled within this octave limitation (Fig. 3.8b).

In practice, the highest recorded frequency on the video tape is still limited by the relative head/tape speed, and it is only typically possible to record about 3MHz of the 5·5MHz transmitted bandwidth on a consumer video recorder. Thus, during recording some of the finer picture detail is lost, although this may be subjectively improved upon replay. However, the vital colour frequencies are located around 4·43MHz, so a separate arrangement is made to record these. The chroma carrier is modulated with another carrier, the lower sideband of which falls somewhere around 700kHz, this then being amplitude modulated and recorded directly on the video tape.

It can thus be seen that the frequency limited colour signals are recorded below the luminance information. Since this luminance is

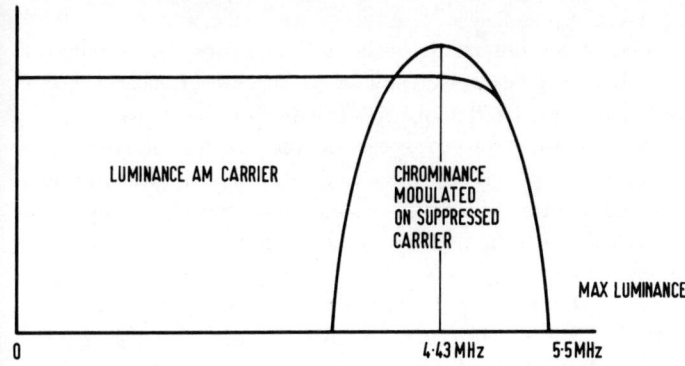

Fig. 3.8a   Television spectrum

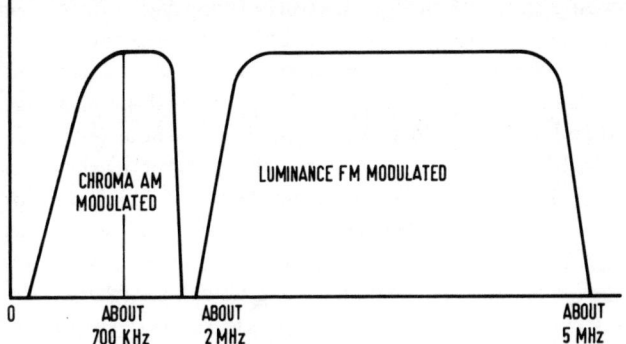

Fig. 3.8b   Recorded spectrum

frequency modulated, it may be recorded directly onto the video tape without any form of bias. Nor does the level being replayed from the tape significantly vary the picture unless drop out (missing tape coating) causes a total loss of signal. Hence in the replay chain, a drop out compensator monitors this replayed level and upon discovering a drop out, replays the previous television line (which it continually stores) so filling the space. Otherwise the replay chain is similar to that of recording with a luminance demodulator and chroma mixer, after which the signals are combined, electronically 'cleaned up' and 'crispened' using high frequency boost to recreate some sharp edges or detail in the picture.

The replayed video and audio are then combined in the RF modulator which is essentially a microminiature television transmitter operating somewhere around channels 35-45 in the UHF band. This modulated output is then combined with the incoming aerial signal

33

and sent to the television receiver where it can be tuned on a spare pre-selector. There are also electronic servo circuits in video cassette recorders which ensure that the head drum spins at a precise speed and which also control the tape speed and physical location of the tape relative to the control track pulses. Accuracy of speed is essential since otherwise the picture will appear to shudder on the TV set and in the worse case, will break up and be lost altogether.

Looking back at Fig. 3.2, it can be seen that between each video track is a guard band that prevents interference in much the same way as crosstalk in audio. However, this is obviously wasteful on video tape, and techniques have been developed which allow this guard band to be eliminated completely and thus provide considerably im-proved recording times. Basically, the two rotating video heads are tilted slightly in opposite directions so that alternative recorded tracks have a different azimuth. If while scanning one track, a video head slightly covers an adjacent track, the level replayed will be consider-ably reduced and thus cause little interference. Three hour recording times would not have been economically possible without this develop-ment, although in America another technique termed 'skip field' recording has been used to double recording times. A television signal comprises 25 frames per second, each made up from two interlaced fields where every other line belongs to one field, thus making the total 50 fields per second. If every other field is 'skipped', recording time is doubled but at the expense of reduced vertical resolution and occasional 'jerkiness' of movement. These penalties are such that skip field is not commercially available in Britain.

Early video tape recorders

The first video tape recorder was developed by the Ampex Corpora-tion (who still hold the basic patents) in 1956 and was designed for broadcasters. The machine itself was very bulky and also required two large equipment bays to house all the complex electronics – it was necessarily extremely expensive. As mentioned earlier, the problem with video tape recording is recording the extremely wide bandwidth produced by the television signal. Although previous attempts had used tape running at high speed, these were unsuccessful and instead Ampex used 2in wide magnetic tape with the video tracks being re-corded virtually at right angles to tape motion, i.e. across the tape rather than along it. These tracks are recorded using four video heads mounted on a rotating head drum, and the speed is such that each head records either 16 or 17 television lines – a number of tracks are therefore required to record each complete picture of 625 lines and this does cause severe electronic problems applying correction to the signals from different heads. This video tape format is however still in use by all television broadcasting companies and is a world standard. Price is however over £50,000 for machines, and so considerable research has been conducted into producing cheaper and more economical video tape formats.

The principles of helical scan recording where each television field is recorded on a single long video track were discussed earlier in this chapter. Sony developed helical recording in the late 1950s and introduced the first transistorised video tape recorder in Japan in 1958 followed in 1963 by the first portable unit. During the 1960s, Sony, Shibaden (now Hitachi), National and several other Japanese companies developed and marketed (not necessarily successfully) a variety of video tape recorders using both ½in and 1in wide tape, all being monochrome. All used open spools of tape and required extremely complex threading procedures to be followed for correct operation. Although a few machines found their way into individuals homes, generally industry was the principal market of these video tape recorders. The restricted recording times of around half an hour also limited their use. Although Ampex in the USA and Philips in the Netherlands both developed and marketed industrial video tape recorders using 1in wide tape, the bulk of research was performed in Japan – Britain has yet to produce an economically successful video recording system of any kind! Both Philips and Sony, aware of the lucrative consumer markets, were fighting a battle to produce the first video cassette recorder, rather than allowing consumers to battle with tape spools and complex threading procedures. Philips and Sony developed different types of video cassette where both tape spools are contained within the casette (similarily to the popular audio Compact cassette) and the video tape is withdrawn automatically from the cassette, after loading, by a small electric motor and some extremely complicated mechanics.

Throughout the history of consumer and industrial video tape recording, there have been a vast number of non-compatible video tape recording formats developed by the various manufacturers. The broadcast format developed by Ampex was immediately taken up by RCA which is the only other company manufacturing such broadcast video tape recorders. Although machines from RCA and Ampex use different technology, tapes produced on both are directly compatible; users have the advantage of not having to rely on a single company for supply of video tape recorders which would obviously cause numerous problems if that company suddenly decided this was no longer a profitable market! Early in the 1960s, a similar problem looked likely to occur with audio cassettes. Philips developed its Compact cassette and eventually provided free licences for all companies interested in manufacturing compatible cassette recorders which have duly appeared in their millions. Unfortunately, no such standardisation has yet occurred with video cassettes.

History shows that on average, each year in the seventies has seen the introduction of a new, and non-compatible video cassette format. Fig. 3.9 attempts to record the introduction of new video cassette recorder formats and new models of recorders where these were particularly innovative in the early years. Immediately it is obvious that the four formats currently competing for market penetration, that is Philips, VCR-LP, JVC VHS, Sony Beta and Grundig SVR were introduced in 1977 and 1978, in fact within about six months of one another. Of the two other formats shown, Philips VCR was the first consumer video cassette recorder marketed in Europe and has now been superseded by the longer VCR-LP format, while Sony U-Matic is primarily commercial and industrial format that provides excellent quality and a wide variety of variants some providing multistandard

Fig. 3.9 History of video cassette recording in Europe

Year	Development
1972	Philips introduce N1500 VCR format (1 hour)
	Sony introduce VO1600 U-Matic format (1 hour)
1973	Sony introduce VO1810 dual standard U-Matic
1975	Sony introduce VO3800 portable U-Matic and VO2850 editing U-Matic
1976	Philips introduce N1501 (early), and N1502 (late) VCR formats
	Sony introduce VO1830 triple standard U-Matic recorder
1977 (late)	Philips introduce N1700 VCR-LP format (2 hours)
1978	JVC introduce HR-3300 VHS format recorder (3 hours)
	Sony introduce SL8000UB Beta format recorder ($3\frac{1}{4}$ hours)
	Grundig introduce SVR4004 SVR format recorder (4 hours)
	Philips introduce $2\frac{1}{2}$ hour cassette for VCR-LP format
	Sony introduce VO2630 triple standard U-Matic (1 hour)
1979	Various companies introduce programmable timers
	Philips introduce N1702 VCR-LP format recorder (3 hours)
	JVC introduces HR-3330 VHS format recorder
	Sony and JVC introduce portable Beta and VHS format recorders
1980	Philips introduce V2000 cassette format offering 2 x 4 hours
1980?	Philips introduce video disc players

operation on the television systems used in different countries, and others designed for programme production with portable and editing video cassette recorders. Although a number of U-Matic format video cassette recorders have found their way into consumer use (primarily in the USA), they are not strictly consumer oriented (price is rather higher than other units) but will be considered later because of their versatility.

Philips first announced its Video Cassette Recorder format (VCR for short) in June 1970, although commercial introduction did not follow until 1972. Just on a point of terminology, since VCR is a Philips registered trade mark, these initials should *never* be used generically as an abbreviation for video cassette recorder. Philips first video cassette recorder using the VCR format was the N1500. The cassette used contained $\frac{1}{2}$in wide chromium dioxide video tape reeled on two spools located one above the other within the cassette package (coaxial design). Threading operated similarly to that shown in Fig. 3.3 and the N1500 included a cooker style mechanical timer that allowed the tape to be started automatically within a 24 hour period. Although designed for consumer use a few hours per week, the early N1500 recorders were used by industry and particularly education, and gained a certain reputation for unreliable operation – this was perhaps unfair since in reality the machines were being used far more heavily than there designers had ever intended in the industry sector, and the chalk laden classroom atmospheres together with maltreatment by inexperienced teachers was also partially responsible for this reputation.

Nevertheless, Philips sold substantial numbers of this model and introduced two later models using the same VCR format, the N1501 (Fig. 3.10) which was merely a mechanically and electrically more reliable version of the N1500 – the N1502 that superseded it less than a year later was a totally new design with modular electronics for simplified servicing, and a completely new mechanical design using a gear driven threading mechanism to replace the 'string and elastic band' in the earlier models. From a user point of view, it also included a digital timer with electronic display which permitted recordings to be set to start recordings up to three days ahead and also allow the recording length to be set enabling more than one programme to be recorded on each one hour long cassette. During the five years that Philips retained a virtual monopoly on consumer video cassette recorders, some 35,000 VCR format recorders were sold, and numerous

Fig. 3.10   Early Philips N1501 with one hour recording time

libraries of video cassettes built up both by organisations so that the
N1502 is still current and used where the economics are such that
these libraries can not be transferred onto the more recent video cassette recorders which offer considerably more economic recording.

Current video cassette recorders

While the Philips VCR series of video cassette recorders provided
excellent recording quality, the principal limiting factor in high consumer penetration was the restricted recording time of 60 minutes for
which a cassette cost about £22. In late 1977, Philips announced the
N1700 video cassette recorder which was basically similar to the existing N1502, but used a slower tape speed to achieve a recording period
of 130 minutes using the same video cassettes that provided only 60
minutes on the N1502. This reduced the cost of recording to about £10
per hour, and subsequently Philips cut the price of these cassettes so
that with typical discounting, recording on the N1700 costs about £6
per hour. The electronic timer has a larger display than the earlier
model, and allows recording to be preset up to three days ahead and
for recording length to be preset. During 1978, Philips introduced a
new video cassette containing additional tape which brings the recording time of the N1700 up to 2½ hours. Unlike the previous N1502,
the N1700 does not provide still frame (a stationary picture, usually

slightly ragged). Finally, during 1979, Philips introduced the N1702 (Fig. 3.11) which is an updated version of the N1700 offering a three hour recording time on a new video cassette (that can also be used on the N1700), a 10 day ahead electronic timer, and one or two other features such as a built-in test signal simplifying installation. The N1700 series was only a 'temporary' measure while the Philips Central Research Laboratories (whose annual budget is almost £500m) developed the V2000 series video cassette recorder which offers four hours either side of a new cassette (Fig. 3.19b).

While Philips has been marketing consumer recorder in Europe, the Japanese have been undertaking further research and development and two new consumer video cassette recorder formats were introduced in Europe during 1978 – Video Home System or VHS from JVC, and Beta from Sony. In each case, numerous other companies are marketing recorders to one or other formats, details of which appear in Fig. 3.12. As can be seen, the JVC VHS format is currently the most popular with some 12 different companies marketing VHS video cassette recorders. What should be made clear is that there are really only two different VHS machines which are actually manufactured by JVC (Fig. 3.13a and b) and National Panasonic (Fig. 3.14). All the remaining recorders are manufactured at the time of writing by JVC in Japan, in some cases with different styling or electronics being added. When the market has developed sufficiently, certain of these other companies (including Thorn in Britain) intend manufacturing in their own plants. Meanwhile, this wide diversity of outlets for VHS format video cassette recorders has meant that considerable market penetration has been achieved. It may be noted that the Baird (Fig. 3.15), Multibroadcast, DER and Granada brands are

Fig. 3.11   Philips N1702 VCR-LP with 3-hour capacity and 10-day timer

Fig. 3.12   Formats used by various companies

SVR	VHS	VCR-LP	Beta (max)
Grundig SVR4004	JVC HR-3300	Philips N1702	Sony SL8000UB
ITT SVR240	Akai VS9300	Pye N1702	Toshiba V-5250
	Ferguson 3239		Sanyo VTC9300
	Nordmende		NEC PVC 2300
	Baird 8900		
	Multibroadcast 8900		
	DER 8900		
	Granada 8900		
	Hitachi VT-3000		
	Mitsubishi HS200		
	National Panasonic NV8600B		
	Sharp VC-5100		

all dedicated to British rental outlets which engage in considerable promotion and have many thousands of outlets which are regularly visited by customers settling their TV rental bills.

Fig. 3.12 only includes brands being marketed in Britain, and there are several other companies engaged in marketing other brands both in Europe and the USA but to include these would only cause confusion. The tape loading system used in VHS format video cassette recorders was described earlier and shown in Fig. 3.4, and this is instrumental in allowing a unit to be produced that is rather more compact than the various other format recorders currently allow. The cassette used resembles an overgrown audio Compact cassette and uses

Fig. 3.13a   JVC HR-3300 VHS format three-hour recorder with one-day timer

Fig. 3.13b   Later JVC HR-3330 which has eight-day timer and simplified switching

½in video tape on two spools mounted side by side. It is thus about the size of a medium thickness paperback book. The JVC HR-3300 VHS video cassette recorder is, in my opinion, the most aesthetically styled unit available (and thus also the Akai, Ferguson, Baird, Multibroadcast, DER, Granada and Hitachi which are identical apart from minor facia changes) since all the operational controls are mounted vertically on the front and are layed out cleanly and sensibly. This arrangement allows the recorder to be located with only about 9in head room (such as a hi-fi cabinet) since the cassette is still inserted through a 'slot' in the top.

Unlike the Philips N1700 which, in its standard form, only has aerial connection, the VHS recorders have in addition auxiliary video and audio inputs and outputs which may be used either for copying between machines, input from a television camera, or a special type of television receiver called a television monitor that accepts a video input directly. Fig. 3.16 shows how a monitor can provide a higher quality picture. Although the video and audio signals are replayed and processed separately, they must be combined in the modulator to produce a signal simulating that of a broadcast television transmitter that may be connected directly to the television receiver aerial socket, from where it passes through the television tuner, IF amplifier and

Fig. 3.14   National Panasonic NV8600B VHS format three-hour recorder

detectors. It can be seen that if outputs directly from the video recorder processing electronics are made available, and if a television receiver/ monitor is used which provides inputs directly to video and audio amps, then the modulator, tuner, IF and detectors may be bypassed with a resulting increase in quality. This can be particularly important if the replayed programme is poor quality initially when a monitor might provide a more stable (less shuddery) picture. Although specialist receiver/monitors or monitors only (less the receiver section and therefore not able to receive TV programmes) cost about double a normal television receiver, there are a number of video dealers that specialise in converting normal television receivers into receiver/ monitors, and these are more economical.

The timer of the JVC HR-3300 VHS is not as comprehensive as that of the Philips and only allows the recording to be started (no

Fig. 3.15 Baird VHS format recorder rented by Radio Rentals

preset length) up to 24 hours ahead. During 1979, the second generation HR-3330 was introduced (Fig. 3.13b) and which featured an
eight-day timer and simplified switching with auto-operation of the
video/TV switch. However, a more complex timer is likely to be
introduced on the new, upmarket JVC HR-6700 VHS recorder which
also includes still frame and variable speed motion. The timer
provides 10 days advance setting, preset recording length, and
most important, full programmable control allowing up to six
different programmes, on any combination of different channels, to

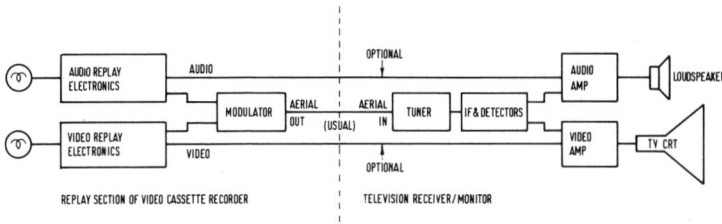

Fig. 3.16 Schematic arrangement for normal method of
connecting recorder and TV via aerial sockets, and
optional video and audio connections

43

be preset for recording. The only limiting factor would of course be three hours maximum recording. The USA HR-6700 has switchable 2-6 hour recording, so 3-6 hour is anticipated for Europe in 1980. New timers of this sort will vastly increase the versatility of recorders allowing you to record first one programme on say BBC 1, then later in the evening, another programme on ITV.

The third video cassette format is Sony Beta which was introduced in Europe during 1978, and is also being marketed by NEC and Toshiba (Fig. 3.17) (actually manufactured by Sony) and Sanyo (which has a more comprehensive timer). Beta uses a video cassette similar to VHS with two spools of ½in wide video tape side-by-side;

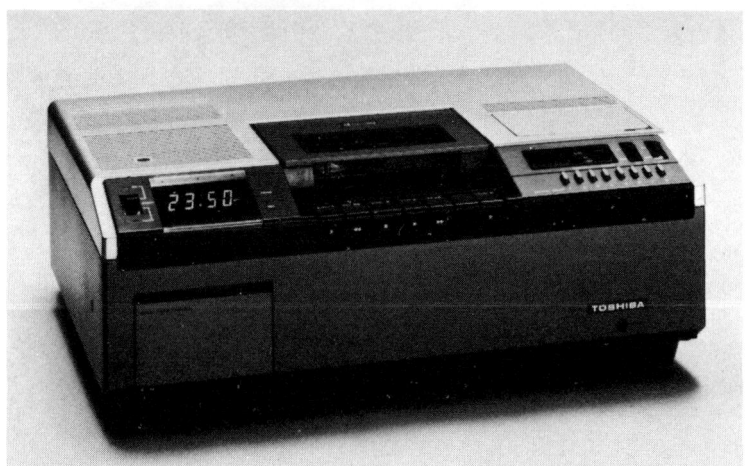

Fig. 3.17   Toshiba V-5250 Beta format 3¼-hour recorder with three-day timer

size is in fact slightly smaller than the VHS cassette. The Sony SL8000UB (Fig. 3.18) Betamax video cassette recorder provides all the usual features but is larger than the VHS machines. The threading mechanism is based on that of the U-Matic format recorders and is similar in principle to that of Philips shown in Fig. 3.3. The timer provides preset recording over a three day period and the recording length can be preset for periods of 15, 30, 45, 60, 75, 90 or 105 minutes or until the tape ends. Maximum recording time on the Beta recorders is 3¼ hours, slightly longer than that of the VHS family. The Sony SL8000UB Betamax also provides video and audio inputs and outputs enabling it to be used with cameras, and receiver monitors, etc. One

Fig. 3.18 Sony SL8000UB Beta 3¼-hour recorder with three-day timer

facility not yet found on the other recorders, is that of still frame combined with the pause function, which may be operated by remote
control. Another facility is that a simple test pattern may be generated
to simplify tuning of the associated television receiver. Recording cost
works out at £4·15 per hour, the cheapest format available.

Grundig has been a licensee of the Philips VCR and VCR-LP video
cassette formats, and during 1978 introduced a new video cassette
recorder, the SVR4004 (Fig. 3.19a) using the SVR (Super Video
Recording) format. Although this uses video cassettes which are
physically identical to that of the Philips VCR format, the cassettes
contain an improved video tape. Thus although SVR cassettes may
be used on VCR-LP recorders, the reverse is not possible and a particular notch on the cassette body prevents this. The Grundig
SVR4004 has two advantages over the other formats; recording time
is currently four hours (with a five hour video cassette rumoured), and
rather than having mechanical operation of recorder functions as have
all the other machines, the SVR4004 control is totally electronic. This
is very important since it is very difficult to remote control functions
such as winding and play when the basic transport uses mechanical
linkages between operating keys and rollers etc., so this feature enables
the Grundig video cassette recorder to provide remote control of stop,

Fig. 3.19a   Grundig SVR 4004 4-hour recorder with 10-day timer

play, record, fast wind, rewind and pause/still frame. Although the timer provided on the SVR4004 will only allow a single programme to be recorded (but preset up to 10 days ahead), the electronic transport control does allow the possibility of programmable operation (as discussed earlier) whereas all the other video cassette recorders would require modified transports to provide this facility. This is because most transports rely on an automatic disengagement of the video tape when the transport is stopped, and only a mechanical movement would allow this to be restarted. ITT also market the Grundig video cassette recorder under the name ITT SVR240. Although recording time is four hours, the actual tape cost is slightly higher than VHS and Beta formats at about £5·40 per hour.

Fig. 3.20 gives a comparison of the four different consumer video cassette formats showing approximate shop prices for the various machines, these being typical selling prices in early 1979, and somewhat lower than the manufacturer's recommended prices. The same applies to video cassette costs which are for the longest recording times available and will be higher for shorter cassettes. It is interesting to note that the Grundig has the fastest rewind time with Philips the least fast. There are several other major differences in operation between the various formats since the video tape must be withdrawn from the cassettes and threaded before play or recording may occur. This threading takes a few seconds, and the different manufacturers have various ideas as to when this should occur. Philips VCR, and

Fig. 3.19b In early 1980, Philips is introducing the new V2000 video cassette system which uses a new Compact video cassette about the same size as a VHS cassette, but which may be turned over at each end, just like a Compact audio cassette. Maximum recording time is four hours on each side, or eight hours in total on a cassette costing about £2·30 per hour (about half existing minimum recording cost). The VR2020 video cassette recorder offers a fully programmable clock/timer allowing five different programmes to be recorded over a 16 day period on up to 26 different television channels. It also includes an electronic digital tape counter that will automatically locate specific programmes on the cassette. Full infra-red remote control is provided.

Fig. 3.20 Comparison of the four different video cassette formats

Format	Max play time (hours)	Rewind/hour (min/hour)	Cost/hour	Tape speed (cm/sec)	Writing Speed (m/sec)
Grundig SVR	4	0·93	£5·40	3·95	8·21
JVC VHS	3	1·3	£4·83	2·34	4·83
Philips VCR-LP	2½	1·8	£6·18	6·56	8·1
Philips V2000	2x4	0·5	£2·30?	2·44	5·08
Sony Betamax	3¼	1·6	£4·15	1·87	6·6

VCR-LP, Grundig SVR and Sony Beta all thread the cassette immediately upon turning on power to the recorder after which all functions are performed with the video tape threaded around the head drum. Although this allows record or play modes to be entered immediately, while the tape is stationary around the rapidly rotating head drum, and if left there for more than a few minutes, will become irreversibly damaged. However, the Beta format threads but has a stationary head drum which needs a few seconds to speed up. So it is not possible to leave these recorders turned on for any period of time, and the fast winding speeds are restricted because the video tape is wrapped around several guides. On the other hand, the JVC VHS family of video cassette recorders only thread the video tape when play or record modes are selected. However, since JVC has developed its new parallel threading mechanism that is slightly faster than the others, this $2\frac{1}{2}$ second delay is not serious, and it has the advantage that winding is faster than Philips and Sony, and that the recorder may be left in standby indefinitely. While in standby, the record key alone may be selected, and the channel tuned on the video cassette recorder, may then be viewed on the television receiver while it is tuned into the recorder – a facility that can often be very convenient.

Video cassettes

Although maximum recording times have been discussed for the various formats, in all cases shorter lengths of tape are available. However, these always work out more expensive per hour than buying the maximum length cassettes. Perhaps an obvious point, but the four formats all use different types of non-compatible cassettes. Nevertheless, there are recognised type numbers. Cassettes suitable for the Philips VCR and VCR-LP formats use the numbers VC30/65, VC45/100, VC60/130 where the first digits are recording time on the earlier VCR format, the larger number being recording time for the VCR-LP (N1700) format. The latest Philips cassettes are labelled LVC120, LVC150, LVC180 these again being playing times on the VCR-LP format. Cassettes for the VHS family of recorders are termed E30, E60, E120, E180 again these being playing times in minutes. Beta cassettes use a labelling system that does not indicate playing time; L125, L250, L500 and L750, having playing times of 30, 65, 130 and 195 minutes, respectively. Finally Grundig SVR format video cassettes are termed SVC and provide four hours recording, potentially to be increased to five hours during 1979.

Although most cassette recorder distributors supply cassettes labelled under their own names, there are also some independent video cassette manufacturers (who also manufacture many of the brand names). Philips VCR and VCR-LP video cassettes are available from Agfa, BASF, Grundig and 3M (Scotch), VHS format cassettes from Ampex, Fuji and TDK, and Beta cassettes from 3M, Fuji and TDK. A comparison of the three different types of video cassette are shown in Fig. 3.21.

Fig. 3.21 Comparison of video cassette sizes, showing Philips VCR-LP (same as Grundig SVR), JVC VHS, and Sony Beta cassettes

Many of us have had experience of damaged audio Compact cassettes. Fortunately, these usually cost less than £1 and can be considered disposable if repair is impossible. With video cassettes costing between £10 and £20, rather more care should be taken. The tape is normally contained within the cassette and is protected by a cover. Never attempt to lift this cover or touch the video tape. Although the cassettes are usually held together with screws, under no account attempt to open the cassette. There are various springs and tags inside that might become dislodged or damaged if the cassette is tampered with. If it breaks, video tape can not be safely joined or spliced since stickiness from the splicing tape might damage the video heads. Occasionally transports misbehave and the video tape becomes jammed within the recorder. Do not attempt to eject the cassette, or remove the tape from the mechanism, but leave it to an engineer. Apart from anything else, an engineer is more likely to be able to remove the cassette with tape unbroken. In practice the video tape will probably have been crumpled and pictures recorded over the crinkles will break up for a few seconds, but this is of minor concern compared with the cost of replacing the cassette.

Occasionally, the recorders automatic stop at the end of the cassette might fail or be overridden by continued depression of the rewind key, and the video tape will become detached from the spool hub. This is a repairable fault which most dealers should be able to manage. However, my experience of attempting to mend cassettes is that although it only takes a few moments to open the cassette (preferably on a large clear surface to catch the bits), it invariably takes over an hour to put the cassette back together! Not a job to be attempted by amateurs. Sometimes, if video cassettes are transported, the ratchet mechanisms can cause the video tape to become stretched and occasionally break. Again, the most likely time for the video tape to become damaged or broken is during loading. For both these reasons, it is always advisable to rewind video cassettes before removing them from the video cassette recorder. Finally, never leave a video cassette recorder in still frame or pause for more than a couple of minutes, otherwise the heads scanning the video tape continuously will cause tape damage.

Maintenance

Video cassette recorders are very expensive and complicated pieces of machinery. They deserve and require careful attention and regular maintenance. The two video heads each scan the video tape 25 times per second and thus suffer heavy wear. Although actual head life varies tremendously between different machines, a typical life is about 1,000 hours of use. Replacement cost varies between models, but is usually somewhere between £50 and £70, plus labour. Generally, the safest principle in looking after one's video cassette recorder is never to touch the interior of the mechanism, and particularly not with any sharp objects. Some manufacturers recommend that certain of the guides are cleaned every 100 hours or so, but check carefully with the instruction manual first. Again, with so many moving parts, video cassette recorders should really be serviced at least once each year to keep them in good condition.

Unlike a television receiver which is primarily pure electronics, the performance of video recorders depends very much on the mechanisms and will deteriorate unless cleaned and oiled at regular intervals (just like a car). Since video cassette recorders are still very new, there is little education in the servicing business so that finding a good video repair engineer can be very difficult. Unfortunately, not all the outlets that sell video recorders have adequate servicing facilities, and many rely on sending units back to the importer's servicing facility, which although undoubtedly providing the service facilities for that particular model, can be time consuming and expensive in transport charges. I would recommend anyone buying a video cassette recorder to check with the shop or outlet about servicing facilities and charges before buying the recorder, and if not satisfied, take your business elsewhere.

The specialist video dealers have a trade association called the Association of Video Dealers who will supply a list of recognised video dealers on request – their members abide by a strict code of conduct. Many dealers offer maintenance contracts, and these are generally to be recommended – video repairs can become very expensive.

Rent or buy?

Before you actually dash out to the shops, first consider whether you actually want to buy a video cassette recorder, or if renting might be advantageous. Depending upon model and discount, present prices vary between £500 and £650 to buy, while renting costs £18 per month. Remember that once purchased, a video cassette recorder is unlikely to have a particularly high secondhand value since new formats and better machines seem to be appearing each year. Also there are a vast number of extremely delicate moving parts inside, and all video cassette recorders require competent servicing (which is difficult to find) at least annually. If the video heads require replacing, a potential bill of £70 to £90 could arrive.

In the UK Philips video cassette recorders can be rented from Visionhire, while VHS units are available from Radio Rentals (Baird), Multibroadcast, DER and Granada for £18 a month, six months advance payment and minimum rental period of one year. This includes six monthly servicing, and one free video cassette. Including servicing, one would probably break even after about five or six years if buying, but on the other hand when a new model with improved facilities is released, one only has to pay another six months advance rental and you can immediately exchange the old unit for new. Worth thinking about?

Advanced video cassette recorders
All the video cassette recorders previously described were designed specifically for consumer use, and as such are primarily for recording television programmes off-air. Volume of sales is such that the manufacturers can produce deviants of the models that are designed specifically for particular countries of the world that use differing television standards. This means that cost savings can be made by omitting certain sections that are not required. For instance there are basically three different VHS format machines in Europe. One for the UK with UHF band only tuner, operating on PAL colour, one for Germany, Scandinavia and Benelux operating again on PAL but with both VHF and UHF band tuners, while the third models operates in France on the SECAM colour signal. A totally different VHS format video cassette recorder is marketed in Japan and the USA. This uses a different tape speed and provides only two hours recording time on full frame, although some models also offer skip field recording (every other field) with reduced quality but four hours recording time. The USA also uses a different colour system (NTSC) and more importantly a different scanning system with 60 field each second and 525 lines (instead of 50 fields and 625 lines in Europe). This means that to provide compatibility between USA and European standards introduces certain compromises which increase the unit cost somewhat.

Before describing the U-Matic video cassette recorder in further detail, perhaps it would be best to examine the problems created by the different television colour standards, and which countries use what.

Colour television was introduced in the USA in 1954 using the NTSC colour system. Because of the technique used for encoding the colour information onto the picture luminance (brightness), NTSC is prone to errors in transmission that cause hue, brightness and saturation errors – thus the tales of green faced people from the early days of colour television. Fortunately, television receivers are better today, but errors still occur. With this in mind, much research was conducted during the early sixties in Europe, in search of a better and more rugged colour system. Two different systems were eventually introduced in 1967 with most of Europe opting for the PAL colour system, apart from France that instead developed the SECAM system. Basically, PAL is very similar to NTSC but reverses the colour coding every other line enabling errors to be cancelled out. SECAM uses a totally different colour technique where the two colour signals are frequency modulated on alternate lines – although the SECAM signal is robust and may be easily transmitted and very simply recorded, it is difficult to process in television studios. NTSC is used in Japan, North America; PAL in Britain, all of Europe except France, Australia, and many parts of Africa, and the Far East; SECAM is used in France, Russia, East Germany and most of the Middle East.

What this all goes to show is that sending video cassettes from one country to another is not quite as simple as it might appear. For instance SECAM video cassettes will only playback in monochrome on PAL recorders, and vice versa. NTSC will not playback at all on PAL or SECAM machines and vice versa. Although a couple of enterprising video dealers have modified VHS and Beta video cassette recorders for triple standard operation (in other words on all three colour standards), these are not generally available and will not all necessarily be compatible with one another.

However over the past five years, the Sony U-Matic format has been developed to provide world wide compatibility with these various television standards. The Sony U-Matic format was introduced in North America in 1972, followed in Britain a year later. It found immediate acceptance in commerce and large libraries of programmes were built up by some of the multinational companies. It was obviously advantageous that these expensively produced programmes could be shown in Europe, so the dual standard U-Matic video cassette recorder was introduced. Fortunately, with the U-Matic format, the linear video tape speed is identical in both PAL and NTSC models, although the head drum speeds are different to allow for the higher field rate on NTSC pictures (60 fields per second instead of 50 in PAL). So the dual standard VO1810 was fitted with a dual speed head drum that allowed NTSC pictures to be replayed onto a special television receiver that would operate with both 625 line 50 field and 525 line 60 field television pictures. It was mentioned earlier that the PAL system uses a 4·43MHz subcarrier to carry the colour information, but the NTSC system uses instead a 3·58MHz subcarrier. However, fortuitously, both subcarriers are too high to record direct onto video tape, and are modulated onto a lower frequency carrier as was shown in Fig. 3.8b. This just happens to be the same for both PAL and NTSC

U-Matic formats so when an NTSC video cassette is played back on a PAL recorder, the NTSC colour subcarrier is shifted to 4.43MHz giving what is sometimes termed 'modified NTSC' or 'NTSC 4.43'. Since the encoding techniques of NTSC and PAL are similar, it is very easy to design a dual standard colour decoder that handles both systems.

With the emergence a few years ago of the Middle East, and the Western customs being slowly introduced, many companies have made substantial profits from marketing video cassette recorders and programmes, in the Gulf area. Unfortunately the colour system used in Iran, Saudi Arabia, and Iraq is SECAM, while most of the programmes to be imported were either American or British in origin with NTSC and PAL colour standards. So the triple standard U-Matic video cassette recorder, the VO1830, was introduced in 1976, and superseded in 1978 by VO2630 and VO2030 (Fig. 3.22) which is play only. Both record and replay PAL, SECAM and modified NTSC and when used with a special triple standard television receiver (Fig. 3.23), can replay U-Matic video cassettes recorded throughout the world.

The Sony U-Matic format (also used by JVC, National Panasonic and NEC) uses $\frac{3}{4}$in wide video tape in a cassette somewhat similar to

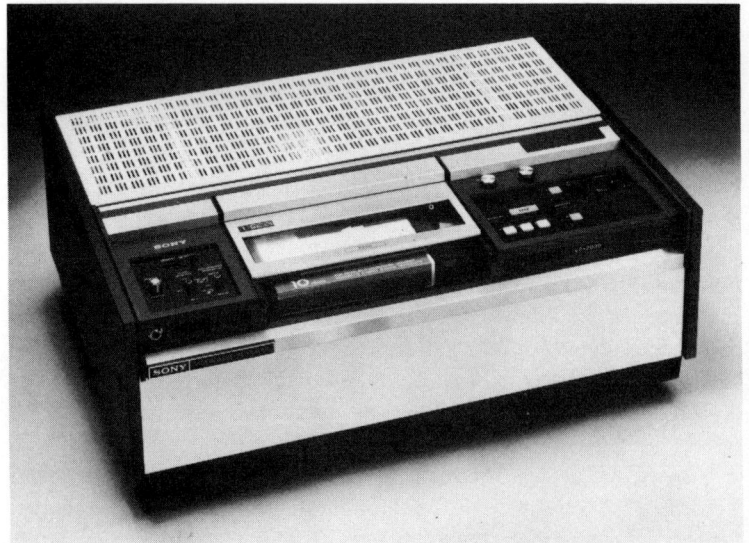

Fig. 3.22   Sony VP2030 U-matic format player, VO2630 is similar

Fig. 3.23 Barco triple standard television receiver/monitor that can operate all over the world on different TV standards

the Beta, but rather larger. Maximum recording time is one hour which costs about £20, about four times more expensive than VHS and Beta per hour. Quality is however rather higher. Unlike the consumer recorders, U-Matic machines do not have a built-in television tuner, nor digital timer, and rely on video input and outputs. To record programme off-air, one can use a receiver/monitor which, in addition to accepting inputs as was seen in Fig. 3.16, also provides video and audio outputs. Prices of the recorders are also higher, the VO2630 costing £1,350 (less tuner and timer). But for many, where programme interchange is necessary, the U-Matic format will be around for a long time yet. Many isolated communities such as oil rigs, ships, overseas construction projects and so on, rely on U-Matic machines to provide entertainment from home. During 1976, Sony introduced two additional U-Matic recorders – the VO3800 (Fig. 3.24) is battery operated and totally portable, while the VO2850 (now superseded by the VO2860) provides full electronic editing facilities enabling proper programmes to be produced using the U-Matic format. In the three years these machines have been around, they have replaced virtually all other video tape recorders for many aspects of video programme production in commerce, industry and education. The VO3800 uses a special U-Matic cassette containing

55

Fig. 3.24 Sony VO3800 portable U-matic format recorder

only 20 minutes of video tape, but which is more compact than the full 60 minute version, and provides 60 minutes recording time on a fully charged battery, while the VO2850 found numerous applications where high quality recording was required and has been used in conjunction with a variety of editing control units for the production of broadcast television news programmes in the USA (electronic news gathering). Some of these programme production aspects are discussed in Chapter 10.

Copying video cassettes and compatibility

Once a small library of video cassettes has been built up, there comes a stage when one often wants to copy video cassettes that other people have recorded. While this is quite possible, it must be remembered that there is considerable quality loss each time a video tape is copied onto another, particularly with these consumer recorders which are primarily designed for one-off off-air recording. Always use the separate video and audio connectors on the recorders, rather than the aerial sockets, since this will enable maximum quality to be maintained. It goes without saying, that to produce a copy of a video cassette, two video cassette recorders are required, one to replay, one to record. It is quite possible to play back off one format, and record onto another totally different format – indeed this is the only way of replaying one type of cassette on another format recorder.

In general, it should be possible to replay video cassettes recorded on one video cassette recorder, on another of the format, but different make or model. In practice, it is possible the tracking control will have to be adjusted to obtain a stable picture on replay. In some cases, it is unfortunately not possible to replay cassettes on all machines – this was particularly so with some of the older video cassette recorders which were not mechanically stable, but the situation is rather better now.

A professional copying bank of VHS format machines is seen in Fig. 3.25, and is where video cassettes are bulk copied.

Into the future
Now that the four consumer video cassettes formats have gained considerable impact, it will be difficult for new formats to be introduced

Fig. 3.25   Duplicating bank of Akai VHS recorders at
Rank Video Centre

with any certainty of success, although the new Philips V2000 system should be a strong competitor if taken up by the rental companies. Both JVC and Sony will be introducing battery portable versions of the VHS and Beta formats respectively, during 1979, and these will be available with low cost colour cameras, the package probably costing about £2,000. Video discs are discussed in Chapter 4, and I believe 1981 will see their introduction in the shops of Europe (although they have been demonstrated here for many years), but it has got off to a very slow start in the USA (due to manufacturing problems) and it now appears that only about 40,000 will be available in 1980; that is less than 5% of the estimated video cassette recorder sales for the year. Thus until Philips/Magnavox can meet the U.S. demand, Europe might have to wait.

4   Video Discs

Adrian Hope

The test launch by Philips in Atlanta, Georgia, of the company's optical video disc system (Fig. 4.1) was a cliff hanger right up to the last moment. For many years now engineers have had a standing joke to the effect that Philips are going to launch their video disc soon. The company first showed the system to the press in late 1972. It was then only in laboratory prototype form but in 1974 Philips formally announced its intention to hold a series of conferences and demonstrations of the system, by then christened Video Long Playing Disc, in Tokyo. 'It is expected that VLP, based on the PAL system, will be available in the UK in 1976/1977', the announcement ended.

As is now history, 1976 and 1977 both came and went without demonstrations, let alone public availability, of the VLP despite continuing enshrinement of the earlier promises in company literature. For a long time it looked as if 1978 would also pass without a commercial launch. But in the summer of 1978 Philips became much more open with detailed information on the working of the system. This release of information was then followed by the promise of a commercial test launch before the end of the year. Finally it was revealed that Atlanta, Georgia in the USA had been chosen by the American Philips subsidiary Magnavox as a reasonably average sized community of reasonably average affluence. The thinking, it seems, was that if Atlanta Georgians could be persuaded to buy a player and video discs of current box office hit films such as *Jaws II*, *Airport '77*, *Saturday Night Fever*, *Sergeant Pepper*, *Car Wash* and *Jesus Christ Superstar* (as owned by MCA, Philips powerful partner in video disc marketing) then the rest of the world might be relied on to follow suit. Vacillation and uncertainty persisted even after the December date of the launch press conference had been fixed. Lack of hard fact information from

59

Fig. 4.1   Philips VLP video disc system launched in early 1979

Philips on the selling price of players and discs and even a nagging feeling in some quarters that the press conference might simply be taken as an opportunity to announce a later launch date, conflicted with the news that players were delivered to some Atlanta shops during November. But it seemed clear that Philips was anxious to go down in history as the company that first sold optical video discs and players in 1978.

On this count, Philips have succeeded. The December Press conference did finally see the Philips system commercially launched, with a price tag of $695 on the player and a catalogue of 200 titles priced between $5·95 for half hour 'How to' flicks, $9·95 for hour long classics and TV movies, $15·95 for the feature films and $20 for musical epics like *The Mikado* and *Swan Lake*. If nothing else, Philips certainly solved the problem of what to give the man who has everything in Georgia last Christmas.

A slow spread of marketing across the USA was planned, with national distribution by the early 1980s. There was talk of sales in the UK 'as early as possible in 1980' but the press, trade and video-aware public will be excused believing this promise only when it comes true.

It is perhaps a significant token of world wide recognition of the potential significance of the Philips system that the press has sustained interest in its progress over such a lengthy period of time and despite so many false starts. No video disc system yet known to man can currently claim to be cheap to manufacture, cheap to maintain, reliable to use and blessed with every technical feature that the user could require. So none is perfect. But the Philips system does appear to come closest to the ideal than any other system developed, announced or launched. And there has been no shortage of developments and announcements. At a recent count the number of systems was put at over 40. And the Philips VLP disc is in fact the third video disc system to be launched commercially, that is to say offered for sale to the public.

The first launch was in the early 1930s when Selfridges sold Shellac discs containing signals coded to produce pictures on the primitive mechanical Baird TV system. Inevitably the Baird video disc died with the Baird TV system. The next commercial launch was in 1975 when Teldec in Germany (a partnership of Telefunken and Decca, thus Tel dec) marketed the TeD video disc system (Fig. 4.2). TeD players and discs were retailed on the Continent for a few years and there is still a medical information service that relies on TeD video discs. But the system was launched too early, both commercially and technically, and foundered. Ironically, laboratory TeD equipment can now produce extremely good video pictures and if the system were to be launched afresh today, who knows, the outcome might be different. But first things first – what is a video disc?

Colour television pictures and sound can be recorded on tape or disc just like sound signals, provided that the recording bandwidth is opened out from the 20kHz or so necessary for sound to the 3MHz or more necessary for colour TV plus sound. Essentially, the only way to extend the bandwidth is to increase the effective writing speed, that is to say the speed at which the tape or disc material passed the playback head. In a video tape system this is usually achieved by running the tape at relatively low speeds (a few inches a second) and spinning the heads past the tape at very high speeds (up to 1500 or 1800 rpm). With a video disc the disc record is spun similarly fast past a stationary playback head. All video discs work on this principle, the various systems relying on different methods of writing the information into or onto the disc and retrieving it on replay. Next, and before discussing and comparing specific video disc systems, it is necessary to digress and remind readers that the whole point of a video disc player is that it should plug into a domestic TV set and replay video discs without modification. However, the Babel-like lack of standardisation between TV transmission systems presents obstacles to this ideal.

Whereas the format for audio cassette tape was agreed around the world in the early 1960s (thanks to clever marketing moves by Philips, inventors of the audio cassette) and the format for microgroove stereo LPs was agreed back in the 1950s (thanks to work by the RIAA in America) there has been little such collaboration in the field of television transmission. This is due mainly to the fact that decisions on TV transmission standards are taken at government level and for political reasons rather than at industry level in the interests of technological progress and to the benefit of the consumer.

There are for example major differences between the colour systems used around the world. Both the USA and Japan have standardised on the NTSC colour system while Europe is split be-

Fig. 4.2   Telefunken-Decca TeD video disc player

tween SECAM (France and the Eastern block countries) and PAL (the rest). In fact not every country uses PAL, SECAM and NTSC in the same way and there are other more subtle differences between even superficially similar transmission systems. Moreover the situation is ever changing, for instance due to shifts in political alignment. At a recent count (1978) there were seven different colour systems based on the three basic formats. Moreover the frame rate, or number of the individual TV pictures displayed on the screen per second, is 30 in the USA and Japan and 25 in Europe. This discrepancy has its roots in the different mains frequencies adopted round the world (60Hz in the USA and most of Japan, and 50Hz in Europe).

For electronic reasons related to frame rate, a video disc designed for use in Japan and the USA must be recorded and replayed at higher speed than the speed for Europe.

With care and forethought on the part of video disc system designers, it should be possible for video disc players and discs to be marketed around the world in just two basic formats; one compatible with domestic TV sets sold for use in NTSC 30 frame/s countries (USA and Japan) and the other compatible with domestic TV sets sold for use in PAL and SECAM 25 frame/s countries (Europe). Switchable circuitry inside the player will 'fine tune' its output to match whatever local standard is required. As things stand at the present, and for the foreseeable future, there is unlikely to be available a single standard player or a single standard disc, compatible at one and the same time with both the NTSC and the PAL/SECAM formats. What this will mean in practice is that European tourists travelling in the USA or Japan will not be able to buy video discs for use on a player back home, and *vice versa*. There will also be no 'import' market of video discs comparable to that which exists for audio LPs.

This worldwide compatibility problem, stemming from the different basic TV transmission formats used round the world, is of course in addition to the quite separate and even more far reaching problem of incompatibility between the different video disc standards either already or yet to be announced and launched. Just as it is impossible to use a VHS video cassette on a Beta machine and vice versa or a Philips video tape on either a VHS or Beta and so on, so it will be impossible to use a video disc of one basic format on a video disc player of another basic format. Using as an example the two modern video formats so far mentioned, it is as impossible to use a Teldec video disc on a Philips video disc player as it is to run an electric oven on gas.

It should now make better sense to examine the different systems already in competition.

TeD

The original Baird disc worked like an ordinary sound recording. This was possible because the Baird TV system was of such low definition and low quality that it required only a very low bandwidth. The TeD system worked (and still works) in much the same way as the Baird system, but with the benefit of complex modern technology. The disc is in fact a round flimsy sheet of foil, punched and embossed from a strip rather than pressed like a conventional audio record. The TeD disc is 8in in diameter, rotates on the turntable at 1500rpm for Europe, or 1800rpm for the USA, and provides 10 minutes relatively high definition colour TV playing time.

The surface of the flimsy Teldec disc (Fig. 4.3) is covered with a very fine pitch spiral groove superficially similar to that found on an ordinary LP record but differing in one major respect. Whereas on an ordinary LP record the groove has both lateral and vertical modulation, that is to say the stylus moves from side to side and up and down, the Teldec disc has only vertical or hill and dale modulation. In fact this exactly replicates the original cylinder sound recording system devised by Edison, which was modulated in hill and dale fashion only. The advantage of vertical modulation is that the groove does not vary in width and this of course makes possible the very tight packing of fine pitched grooves. This in turn enables a reasonable length of video playing time to be crammed onto one side of a disc, even when it is rotating at 1500 or 1800rpm.

The colour TV picture and stereo sound information is recorded in the Teldec groove by frequency modulation (as used to transmit stereo radio in all countries around the world) rather than amplitude modulation (as used on an ordinary audio LP and for medium and long wave radio transmission). A tiny combined stylus and ceramic pick-up transducer is secured to the end of a delicate cantilever (Fig. 4.4) and tracks the Teldec disc as it spins supported on a thin cushion of air, Fig. 4.5. The transducer converts the hill and dale modulation into tiny electrical signals, of which the frequency modulations are decoded to produce colour TV and stereo sound signals at an output socket of the player. A connecting lead is plugged between this and the aerial socket of a TV set. Original versions of the Teldec player suffered from various faults, such as poor picture quality. Also some

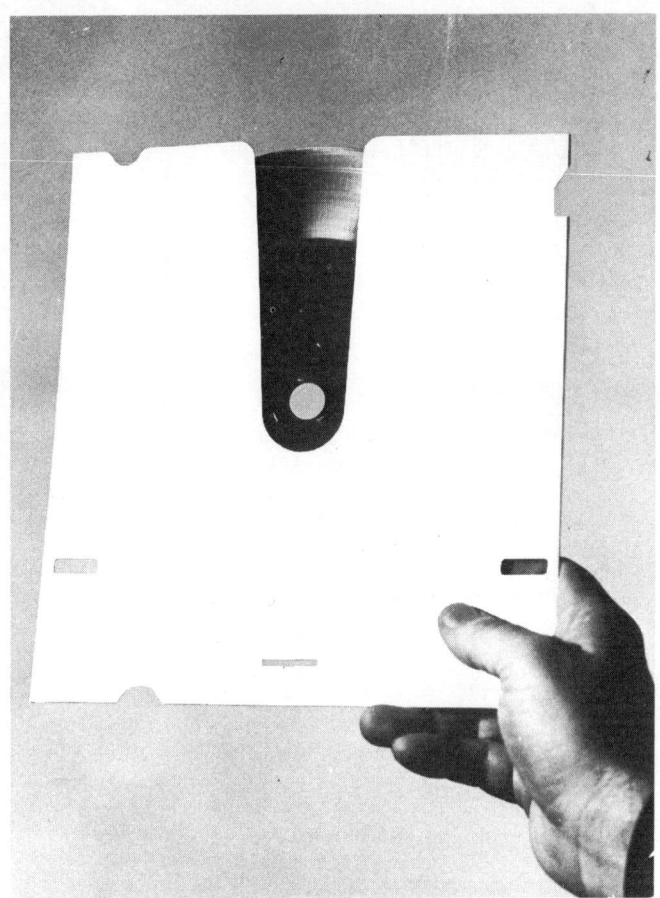

Fig. 4.3   TeD mechanical video disc in cardboard sleeve

users were unhappy with the auto loading system. The flexible foil is contained in a protective sleeve and when this sleeve is loaded into the player a mechanical device automatically removes the foil from the sleeve for playing and thereafter replaces it. Early machines proved somewhat unreliable in the respect. Also the 10 minute maximum playing time per disc was thought by many users to be inadequate. On the credit side, the flexible disc is easily sent by post and can even be incorporated in a magazine as a give-away. Albeit in low profile, the Teldec designs are still continuing work on the system to extend playing time and improve reliability and picture quality. Without doubt

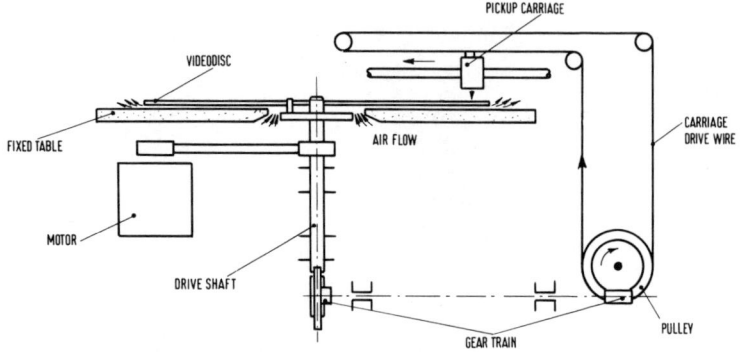

Fig. 4.4 TeD player schematic

the pictures now available from a TeD disc player are extremely good
and it seems likely that if the system were launched afresh today, or
actively relaunched at a commercially ripe time, it might find favour.
Perhaps the main advantage is that, being essentially mechanical, the
system is relatively simple and easy to manufacture and the flexible
discs can be cheaply mass produced.

Visc

One problem with a Teldec relaunch would be the 1978 demonstration by Matsushita (National Panasonic) in Japan of the Visc system.

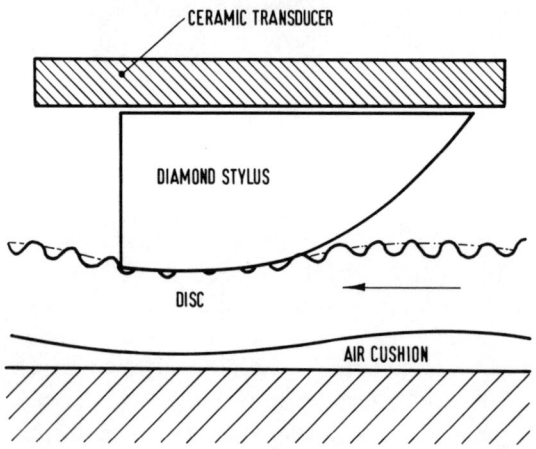

Fig. 4.5 TeD pressure contact pickup

66

In many respects Visc took over from where TeD left off. Like TeD, Visc is a mechanical system and also like TeD, Visc uses an 'ordinary' disc pressed on conventional pressing machines and Matsushita has managed to cram as much as one hour of colour TV and stereo sound on each side of the disc, giving a total of two hours per album. Results as demonstrated in Japan were certainly impressive and it would probably be hard, without direct side-by-side comparison of the same programme to choose between a TeD picture and a Visc picture on the screen. How have Matsushita managed to cram so much more video information on a Visc than Teldec have managed to cram on a TeD disc? One answer of course is that Matsushita are using a 12in rigid vinyl LP format rather than a flexible 8in foil format and thereby sacrificing the ease of postage and magazine-give-away facility offered by TeD. Matsushita have also slowed the rotational speed. Whereas TeD runs at 1500 or 1800rpm, Visc runs at around 450 or 500rpm (as usual dependent on the country for which it is intended). For obvious reasons this increases playing time but it also makes less practical one facility offered by TeD and not offered by Visc. That is stop motion or freeze frame and instant replay of a short selection of the programme. When a video disc rotates at 1500 or 1800rpm, it contains one full picture frame of information per revolution. Thus, halting movement of the replay head across the disc surface to hold it fixed over just one revolution of the groove ensures that the same picture is repeated over and over again to produce a still or freeze frame. If the replay head is skipped back a few grooves, a few frames will be repeated over and over again. According to the lower speed Visc system there is more than one picture frame of information for each revolution of the disc. As a result it is difficult to provide for freeze frame operation and instant replay of a few seconds of action without picture break-up.

Both Visc, and TeD in its 1978 laboratory format produce good picture results from equipment, which being essentially mechanical, is fairly cheap to produce. Both use readily mass-produced discs. Visc, especially, is a technological miracle, but neither system is likely to succeed in the long term because they both rely on the hundred year old and essentially primitive system of tracking a groove with a stylus. Once the disc groove is damaged, the disc is effectively unplayable. Indeed, for this very reason, it seems likely that Matsushita will not pursue the commercial development of Visc as a carrier for video disc programmes. Possibly the company will explore the use of Visc as a carrier for digital sound. (Sound recordings made in digital code provide superb hi-fidelity but require far greater bandwidth than is available from an ordinary audio LP. Thus they must be recorded on a high density carrier such as a video tape or video disc.) Another contributory factor to the likely demise of Visc as a video disc system is the announcement by JVC, a sister company to Matsushita who developed Visc, of a quite different video disc format working along wholly different lines. This situation is politically interesting because it parallels that which occurred a couple of years before Visc, when both JVC and Matsushita developed their own video tape recording formats and Matsushita backed down in favour of the JVC VHS format to which it is now committed.

Selectavision and JVC

The new JVC video disc system (Fig. 4.6) and its technical advantages, are best understood following consideration of a system developed by RCA in the USA and tentatively announced several years ago at around the same time as the Philips optical system. The RCA system,

Fig. 4.6   JVC capacitance video disc player which takes a slot-in video disc in a protective cover

called Selectavision, is part mechanical, part electrical in concept. A rigid plastic disc like a Visc has a groove, like the TeD or Visc groove, and is tracked by stylus. But according to the RCA system the spiral groove tracking is used only as a guidance for replay, not as the actual means of replay. Instead of sensing the groove modulations mechanically, the RCA stylus tracks mechanically but senses electrical variations in the groove. This sensing is by capacitance. The disc surface is conductive in the manner of one plate of a capacitor. The other plate of the capacitor is formed by a metal layer on the tracking stylus. The conductive groove is pitted so that there is a continual variation in capacitance between the stylus electrode and the disc. These changes in capacitance are used to modulate a tuned circuit, and the tuned circuit modulations are decoded to produce colour TV and sound signals. The RCA disc has never yet been commercially launched as RCA, unlike Philips (with MCA) has no ties with a company holding a vault of highly saleable programme material. However, the RCA disc is under continual improvement and development and can now offer one hour per side of a double sided disc. Current designs of the RCA system also offer facilities hitherto believed impossible for a grooved system rotating at any speed less than 1500/1800rpm. Traditionally this rotation speed, where one frame of the picture is recorded during one turn of the disc, has been regarded as essential to the freeze frame display of still pictures when the movement of the replay head across the disc is stopped to replay the same picture over and over again. It has also been assumed that the delicate stylus would not take kindly to being skidded backwards and forwards over the disc to replay a short passage over and over again or locate a selected still picture or passage at the operator's will. Both Teldec and RCA have proved that their system stylus, which in each case tracks very lightly, can perfectly well be skidded backwards and forwards. Also, by the use of complex modern solid state memory circuits, RCA can now provide repeat action and freeze frame while maintaining the rotation speed at the previously proposed low rate of around 500rpm (depending on the country). One advantage of the RCA system is that (like Visc) the records can be stamped out on ordinary audio presses, the presses simply being loaded with a special conductive plastics mix when required to press RCA video discs.

RCA discs like Visc must however be pressed very flat to be usable and be carefully handled because once the very fine grooves on the surface are damaged, the stylus will no longer track accurately and

69

the disc will be useless. RCA have designed a special caddy loader which provides automatic disc loading and ensures that the user never, ever touches a disc surface. This resembles the Teldec system protective sleeve which ensures that the foil disc surface remains untouched by human hand at all times. Because the RCA system (like TeD and Visc) is essentially mechanical and relies on technology developed over many years for audio disc reproduction, suitable players can be constructed from simple and often existing components. This gives all the mechanical systems a price advantage over more complex systems. But inevitably such a price advantage will tend to be eroded if a more complex system moves into high gear mass production.

Incidentally, the stylus of a mechanical system, which must be very finely contoured to trace the tiny grooves of a disc surface, will inevitably eventually wear out and require replacement. RCA has already tackled this problem head-on by designing a drop-in stylus unit that quite literally can be changed by the user in under 10 seconds. Demonstrations of the RCA system in New York have showed it to be capable of producing excellent pictures on screen (as indeed have recent demonstrations of the Teldec, Visc and Philips systems). Nevertheless, there has always been, and still is, a great deal of uncertainty in the video trade over the system future and no one in RCA is giving any clues. It seems probable that, like so many other firms with systems developed in the laboratory, RCA is waiting to evaluate the market for other firm's video discs before a commercial launch of their own technology. This may or may not prove to be a wise decision. Teldec have clearly suffered from launching too early and it may be that others will suffer by launching too late.

Certainly the RCA Selectavision format has already taken a knock from the recent announcement by JVC of its own version of the capacitance system. The long term advantage of the JVC system over Selectavision is that it dispenses altogether with grooves on the disc surface. The short term disadvantage is that to dispense with grooves requires a complex electronic control system. This will inevitably inflate the price until mass production is under way, unless the initial price is heavily subsidised by the backing companies. According to the new JVC capacitance system, the disc surface is conductive but smooth (Fig. 4.7) i.e. without grooves, and covered with a spiral of capacitance pits generally similar to those found in the grooves of the RCA disc. On the JVC disc the pits which carry the programme information run alongside a similar series of pits which are used to

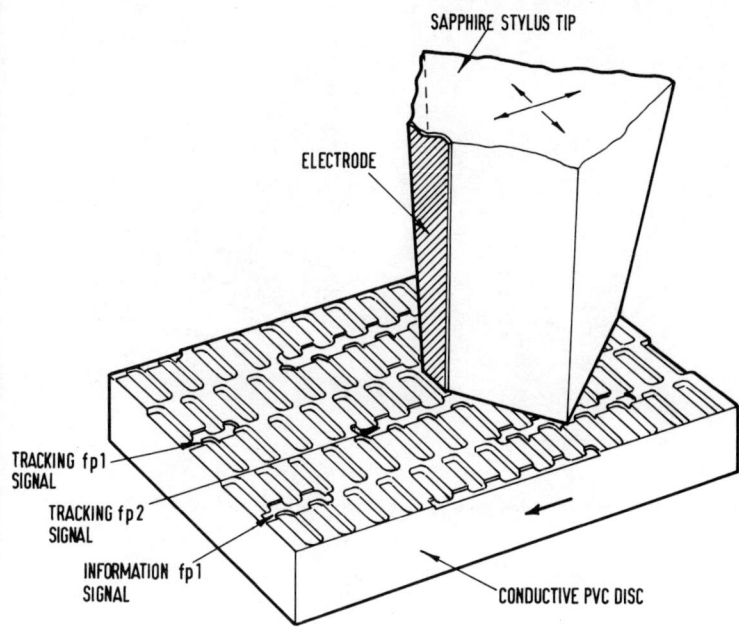

Fig. 4.7   JVC capacitance disc with sapphire stylus

control a servo circuit which guides the electrode stylus. Thus there is no mechanical guidance of the electrode stylus by the disc, other than straightforward vertical support and there is less stylus wear. It seems on the cards that an alliance between JVC and RCA may be necessary, even if not particularly desirable for the parties involved. The two companies do however already have a history of co-operation, having both backed the now virtually obsolete CD4 quadraphonic system.

Such conjecture, along with conjecture on what other new systems are in the pipeline and which of the systems already developed will be dropped or commercially launched, is likely to be resolved once the commercial results of Philips long awaited test launch are known. In short, if the test launch by Philips proves a commercial success, this could well trigger the launch of all other technically viable systems and the merger and modification of similar but not identical interests.

VLP

Without doubt the Philips system (Fig. 4.8) has much to offer. A disc (the same size as a Visc or RCA disc) is coated with a reflective silvery

Fig. 4.8 Philips VLP video disc player with still frame

surface (Fig. 4.9) which is pitted along a spiral track. The spiral track of pits is 'read' by a purely optical pickup which directs a tiny laser beam down onto the disc surface and at the same time senses the reflections coming back from the disc (Fig. 4.10). The pits modulate the reflected beam and the reflected light sensor produces a modulated signal which can be decoded to provide colour TV and sound signals. There is no physical contact between the optical tracking head and the disc surface, the head being moved across the disc surface to follow the spiral of pits by a servo control. There are two versions of the Philips disc system, one which rotates the disc at 1500/1800rpm and another which rotates the disc at a continually varying speed of around 500rpm. According to the first speed option (both are provided for on a single player), there is of course a ready facility for stop motion or repeat action. According to the second option, there is longer playing time but no ready possibility of stop motion or repeat action. But as all those firms who have abandoned stop motion and repeat action will readily point out, anyone watching a feature film on video disc is unlikely to want stop motion or repeat action. It is mainly used for educational programme material.

The reference above to the second option Philips system relying on a rotational speed which *varies* around 500rpm may seem odd but was deliberate. The *linear velocity* of the disc past the playback head is kept constant by varying the rotational speed of the disc as it plays. Remember that on any gramophone record with a spiral groove track, the playback head travels further and faster at the large circumference outer grooves than it does at the small circumference inner grooves. This is wasteful of playing time and by keeping the linear speed or tangential velocity constant throughout by varying the rotational speed as the disc plays there can be a great saving in playing time. The idea was first tried out way back in the 1920s with ordinary gramophone records but failed to catch on because it was too difficult to achieve with the technology then available. Using an optical system with lasers and highly complex electronic computer-style circuitry it is relatively easy. This is what Philips have done, providing the two options at the flick of a switch or even automatically sensed, depending on what kind of disc is to be played – constant rotational speed

Fig. 4.9 Philips reflective video disc

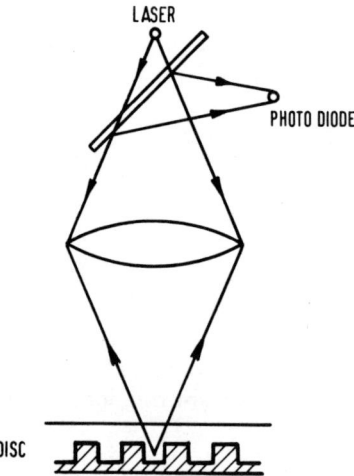

LASER

PHOTO DIODE

DISC

Fig. 4.10 Philips VLP optical
pickup with laser and photo diode

discs for education, etc. (where freeze frame and replay is needed) and constant velocity discs for feature films and the like (where the most important thing is to cram as much playing time on each side of the disc as possible).

Several other firms (including Mitsubishi, Teac and Sony) have generally similar optical systems under development and hopefully some kind of agreement can be reached with cross-licensing for standardisation and compatibility in the future. Pioneer has in fact already adopted the Philips approach. An American firm I.O. Metrics was at one time working on a modification of the Philips approach, whereby a disc of ordinary photographic material is scanned by an ordinary white light source rather than a laser. The French firm Thompson-CSF has independently developed an optical video disc which closely resembles the Philips system except that instead of being reflective it is transmissive. The disc can be made out of flexible and transparent plastic material rather than solid and reflective plastics. The laser beam is directed through the disc and modulated by variations in its optical density. Very probably players can if necessary be designed to cope with both transmissive and reflective types of disc.

The main disadvantage of the optical systems is that they rely on gas lasers which are currently expensive. They also need complicated servo equipment to keep the playback head tracking accurately with the pits always sharply in focus. Philips appear to have cured these problems but at least initially optical players must cost more than mechanical players, or be heavily subsidised by their manufacturers.

MDR

All the video disc systems so far mentioned have one thing in common – they are capable only of *replaying* recordings and are not able to *record* material for subsequent replay. Technically it would not be too difficult to produce a domestic video disc system which offers both a record and replay facility. For several years now TV stations have used magnetic video disc recorders which continually record the signal being transmitted from a sports event and provide for instant replay or freeze frame of chosen passages at the will of the commentator. The disc machine is simply switched from record to replay, and the magnetic head moved back over the disc surface by the required amount. There has already been some research work on a domestic version by MDR in West Germany (Fig. 4.11). Likewise developments made independently by both Philips and Matsushita have shown that it is perfectly feasible for an optical video disc to be user-recorded by providing a high power laser in the machine which is capable of burning the necessary optical pits into the disc surface. The disc is then ready for instant replay. There is even a possibility that

Fig. 4.11   MDR experimental magnetic disc recorder developed in West Germany

the burns may be made temporary, i.e. repairable to make the instant replay recording erasable. Engineers at Thompson-CSF are also known to be working on a record facility for the transmissive disc system. But although an instant record/replay magnetic disc system is of indisputable value for a broadcast TV station, for instance to instant freeze or replay a crucial soccer goal or boxing knock-out, and although optical record/replay systems are of likely value for business data storage in video fiche fashion, there seems little or no point in offering the public a domestic video disc system with record as well as replay facilities. The record facility is available already in all the video cassette recorders in the shops and it is now generally and widely acknowledged that the markets of video tape recording and video disc reproduction are as different as the markets of audio tape recording and LP disc reproduction.

Conclusions

The well-heeled home of the future will boast both a video tape recorder and a video disc player. It may well be that the video disc player will double as a reproducer for digital sound recordings. That is in fact another long and tortuous story, because as yet there is no agreement between the major manufacturers on the digital coding standards to be adopted. Also there are two schools of thought on the desirability or otherwise of what can conveniently be called a hierarchical approach. According to such an approach a single disc player is capable of reproducing video discs through a TV set or digital sound recordings through a decoder, both types of recording being sold on a similarly styled disc. Already Philips, by announcing development of a compact digital sound disc player for the reproduction of digital sound discs which are quite different in size and appearance from Philips video discs, has publicly eschewed the hierarchical approach.

Even setting aside the issue of video disc and digital sound compatibility, it seems that there are many, long and bitter commercial battles ahead. So many firms have now spent so much time and money on developing video disc systems that it is inconceivable that any one firm with a fully developed system will simply back down and throw in the towel in the face of competition from rival systems. The more rival systems that continue to struggle for the market, the less chance there will be of standardisation on any one system and the less chance of anyone achieving commercial success. The crucial point is that no domestic video disc is any use without video disc material available to play on it. However no programme manufacturer will put out video discs unless there are a fair number of players available and thus a ready market for their discs. It is the old chicken and egg vicious circle once again. The last circle was the one that killed off quadraphonics. There is also the quite separate question of who and how many people will buy feature films on video disc format to watch at home anyway. No one knows the answer to this question but it must largely depend on the price at which feature films will be sold. This creates another chicken and egg situation because the only way to sell films on video disc cheaply will be to mass produce and mass market them; and there is no point in mass producing something that only a few people are equipped to play. Without doubt there will be a great deal of loss leader marketing early in the video disc struggle ahead. In fact it has already started. The prices set for Philips Magnavox disc players and disc versions of MCA's hot box office hit films clearly suggest that both Philips and MCA are heavily subsidising their opitcal system according to a 'jam tomorrow' philosophy. It follows that their competitors in the field will be forced into a similar marketing approach. For a few years at least the public will be able to buy players and discs at attractively low loss leader prices. Hopefully, by the time any one system prevails and becomes a world standard, savings accruing from mass production on that system and royalty agreements with the performance unions based on volume sales, will be sufficient to maintain the price of players and programmes at reasonably low level. Those companies which go to the wall and are forced to write off their own systems and cash investment, may well rue the day that they ever put money into video discs. So, I imagine will any member of the public who buys equipment for a system that subsequently goes to the wall. The best they can then do is hold on to whatever becomes obsolete and hope that it becomes a valuable antique in its own lifetime, like an original Edison phonograph.

5 Videograms – Programming for the Home

Angus Robertson

Films, programmes, programs, software, movies or whatever – a variety of names attempting to describe the same media, that of moving pictures recorded for posterity. Call them what you like, but the final meaning is the same – entertainment. Videograms is the latest generic title attempting to encompass the distribution of 'programmes' into the home environment. Similarly, the phonogram was the term initially used to describe music distributed to the home market, although 'record' has become a more acceptable title. Over the years, videogram will undoubtedly be superseded by another name, but temporarily at least, that is the name we must use.

Distribution of videograms has only just begun in the UK, although it has been building up for two years in the USA. The reason for the lack of programmes in Britain is simply that, as yet, there are very few homes capable of replaying videograms because it is necessary to possess a replay machine – either a film projector, video cassette recorder or video disc player. Although 16mm film projectors have been available for around £400 for decades, remember that is somewhat less than the current price of a video cassette recorder and about the same as a video disc player (which has not as yet been launched in Britain). So why have 'films' not been distributed more widely in consumer markets? Fig. 5.1 gives some idea of the answer and shows the raw costs with sale and hire prices for a typical two hour feature movie. A 35mm theatre film print is used in commercial cinemas and is the basis of virtually all feature movies made and distributed around the world. All other distribution media are usually copies of 35mm theatre prints (apart from occasional programmes originated on video tape using television techniques). Not only are 35mm prints expensive, but they are very bulky, being some 10,000 feet of film con-

Fig. 5.1 Comparison between costs of different programme distribution media for a two hour film

Programme	Raw cost	Sale	1-day rental
33mm theatre print	£500	—	—
16mm film print	£200	—	£18
U-Matic video cassette	£75	£100	—
VHS/Beta/VCR-LP video cassette	£15	£35	£6
Video disc	£1	£9	—

tained on five spools each around 15in in diameter and 2in thick. A 16mm film print costs somewhat less than a 35mm film print and is less bulky, a two hour film being 4,300 feet contained on three spools but only 1in or so wide. Sending such a film through the post is difficult, and most 16mm film prints are distributed by rail.

There are vast libraries of 16mm feature movies (and shorts) available from film libraries, and these are primarily used in 'shut-in' locations such as boarding schools, hospitals and such institutions, film clubs, ships, and small cinemas. The typical cost of such films (which are normally only available for hire) is £10–£25 per day, depending upon the age of the film. However film prints are 'mechanically' – or more correctly chemically – produced. They can be easily damaged by careless projectionists or faulty projectors, and old films have a nasty habit of breaking during important shows. Even broadcasting organisations who use the best equipment do not always trust film, and often transfer film onto video tape before broadcast.

However, 16mm libraries are not only available to commercial and educational users, but can be hired by any individual prepared to handle the bulky package and pay the hire charge. Of course the film requires a noisy projector (however cheap) which might be acceptable in a hall, but rarely in the home. The introduction of video cassettes created a new distribution medium with considerably reduced duplicating (raw) costs. I have included U-Matic cassettes here although not used widely by consumers in Britain. Nevertheless there are already large numbers of films available on this commercial format for use in hotels (with automated video cassette players offering movies 24 hours per day on the TV system) and players on ships and remote industrial 'colonies' such as oil rigs, construction sites and so on who want to keep in touch with 'civilisation'. But U-Matic cassettes only have a one hour playing time, and so at least two are usually required for feature movies.

However, during 1978, four new video cassette formats became available in Britain and Europe: JVC VHS, Sony Beta, Philips VCR-LP, and Grundig SVR. The first three offered around three hours, and SVR four hours playing time. This was very important for videograms since it meant that feature films could now be recorded on a single video cassette and this provided the break point in cost; but then of course it was also this new generation of video cassette recorder that was successfully marketed to consumers. Thus from an estimated 25,000 video cassette recorder owning homes at the beginning of 1978, some 100,000 units were estimated to have been placed by early 1979, rising to over a quarter of a million by 1980. If you think that the present record market is only buoyant with a potential market place of over 30 million record players, and that is with heavy radio promotion. Obviously it will take many years before the selection of videograms rivals the choice of records presently available from tens of thousands of outlets.

However, by that time it will not be the video cassette that achieves high market penetration for videogram sales – it will be video discs – and Fig. 5.1 shows precisely how because the manufacturing cost is so low that potential costs are not unsimilar to double albums of today. The reason for this staggering reduction in cost is due to the 'duplication' process used for the different systems. Video cassettes have essentially to be recorded in real time, i.e. duplication comprises banks of video cassette recorders being fed from a master copy of the film or programme, each taking one copy in however long a time the programme lasts. Presently, the market demand can be satisfied by producing only about 6 copies at a time; but as market size grows, this will rapidly increase so that one London duplication facility intends to install over 200 slave machines taking copies from one master (these are approximately in the region of 100 VHS, 60 U-Matic, 50 Beta and 25 VCR-LP – their opinion of the various market demands). Obviously duplication costs come tumbling down as quantity increases, not only due to this multiple copying, but also because the cost of preparing the master (new film print, sound equalisation to match TV sets, copying onto video tape, etc.) are recovered as many copies are run off. Colour printing on the packaging (one expects high quality boxes and printing when spending over £40) is just as important, and again preparation might be a few hundred pounds which is recovered in low unit cost.

National Panasonic developed a high speed video cassette (or

Fig. 5.2   Cassette box illustration from Intervision

cartridge as it was then) duplicator a few years ago, and this is being resurrected for the duplication of VHS video cassettes. Basically, it uses a special metal tape as the master, and this is passed through heated rollers in contact with the raw tape, the heat transferring the magnetic 'image' onto the raw tape much like print through on audio tape where a loud sound can become impressed on tape on adjacent turns. However, cassettes are still mechanically complex and are unlikely to drop in price much below £10 each, so with duplication costs, royalties, copyright, promotion, packaging, overheads, distribution costs and profit margins, videogram cassettes are always going to be at least £30. In audio records, manufacturing only comprises 13% of final product cost, a substantial 58% being promotion, distribution and profit.

This is of course where video discs will offer considerably better economics, since once the master has been produced, it can effectively be 'stamped' out in large numbers at very low unit cost. Such manufacturing is only economic in large quantities, 5,000 being suggested as the smallest run that might be economically produced. Due to the different scanning systems between the USA and Europe, video discs that are made available on the USA market, will not be playable on standard video disc players in the European market. On the other hand, two different sound tracks can be offered on the same disc, so multi-language copies be produced for different European markets.

As described in the previous chapter, video discs have just been launched in the USA and some 200 different video discs sets (many videograms covering more than one disc) are becoming available, initially ranging from shorts at $5·95, to regular feature movies at $15·95 and some long operas at $20. Many of the films are first release, and some have not even been released in Britain as yet. So there is currently an argument (with which the author totally agrees) saying that one should be purchasing USA standard video disc players and video discs, and using either a special American TV, or a dual standard TV to enable these American discs to be seen in Britain. Provided one has a business contact travelling to the USA regularly, the large market there is going to have a substantially better choice of videograms available somewhat earlier than the smaller European market, say until 1985.

Will video disc libraries be set-up? Well, consider books for a moment. There are presently thousands of libraries in Britain, all with vast selections of books available on free loan. Those libraries offering records, are generally making a charge and almost universally have a poor selection of 'modern' music. Thus record libraries are not particularly well patronised. Will these libraries consider setting-up videogram libraries as well, when the record libraries seem to be more trouble than their worth? It is true that some British companies such as Intervision and Radio Rentals are offering video cassettes for daily hire – features films from the former, general interest shorts from the latter. On the other hand, unlike records, video discs are virtually immune from damage and so are a rather more reliable rentable medium than scratched records!

So from Fig. 5.1 it is obvious that the video disc is the cheapest distribution medium and all others are probably rather more expensive than is conducive to the consumer market at present. Although £9 might be acceptable, £35 is certainly a lot of money for a single film. The £9 could be recovered by only say six visits to see a cinema performance (an average family outing including transport), while the £35 would require numerous visits to recover the money. Thus although video discs are obviously the distribution medium of the future, in the interim video cassettes videograms are being sold to those who can afford such luxuries, while some companies realising these market conditions are offering video cassettes for daily rental, generally at around £6 for a feature film. Although there is a possibility that such video cassettes might become available on loan from lending libraries (or for minimal rental as with records), no plans have yet been announced and with the cut backs in library budgets, so this does seem to be a rather remote possibility. However, a large number of specialist video dealers are now holding small libraries of video cassette programmes, and one national television rental organisation is offering video cassettes for daily rental, although only non-feature material at the time of writing.

You would be right to think that there are hundreds of thousands of hours worth of feature films, television programmes and shorts available for distribution as videograms, but due to copyright, only a fraction of that material is likely to become available in the near future. This is very sad since I can personally recall many television programmes that I would welcome the opportunity to see again, many programmes that might only have warranted minor interest and not worthy of television repeats but which will otherwise die a death gathering dust in film and video tape vaults. When the programmes and feature films were originally made, the agreements and contracts negotiated with the various actors, scriptwriters and production team only covered (typically) three television performances or theatrical release, with an exception for television transmission of feature films five years after original release. To release videogram copies of such programmes requires that new contracts are negotiated with the original companies and people involved in the particular production in question. Since this must be accomplished for each and every production, naturally it will become uneconomic to negotiate the release of copyright for minority interest programmes, although production companies are becoming aware of this new distribution medium, and many current agreements being negotiated now specifically make allowance for videograms.

It would also be possible to produce agreements or legislation that would open up this vast library of copyrighted archive programme material to videogram distribution, and this could mean substantial additional fees for those working on the original productions. For instance, the record business in Britain is currently worth over £2bn a year which brings in over £400m for the original copyright owners and artists, not including fees for the product being played on radio or used in public entertainment.

So, for the moment, forget any chance of buying last year's top Christmas TV Show, although some broadcasting companies such as the BBC offer documentary and specialised feature material for sale or hire since this only usually requires the clearance of copyright with staff rather than actors. Nevertheless, feature films can be purchased from EMI Videogram Productions who offer a number of old Ealing comedies with a smattering of more recent films. However, only a minute fraction of those are in the EMI 16mm film rental catalogue. Intervision offer over 200 feature films on sale for around £40 or rental at £5·95 for two days, although it must be said that the majority were not exactly box office successes while on theatrical release. Primarily aimed at clubs and discos are a range of music programmes, many from West Germany where copyright is not quite such a problem as Britain. VCL Video Services is another British company offering videograms on video cassette formats, and their catalogue includes a few tens of feature films, music programmes, sport and some television entertainment.

Although one might expect publishers to become involved in this new medium, so far only IPC Video has released a range of videograms covering leisure aspects such as motorsport, tennis, fishing, skiing, sailing, horses, golf and birds and these are available for sale at around £35 each, or rental through Radio Rentals at £3 for the first day, £2 for the second.

Other types of videograms expected to become available in the coming years include coverage of many other household areas, hobbies and leisure interests such as DIY, gardening, child care, woodwork, photography, gymnastics, animals, fashion etc. Formal education, such as Open University programmes, is another possibility, releasing students from attending their televisions at unusual hours when programmes are currently transmitted. There are many other areas of entertainment too, such as converts, cartoons, history, etc. plus reference programmes that could be considered as visual encyclopaedias, and advertising videograms such as mail order catalogues and sponsored programmes about cars and so on produced by manufacturers hoping to sell the promoted products.

In the USA, there are now over 400,000 video cassette recorders and so the market for videograms is somewhat larger. Two areas predominate; hard core pornography and films for which the copyright has expired or for various legal reasons is invalid. Unless renewed, copyright expires after 30 years and so there are numerous films dating from before 1948 available for direct sale, many of which are particularly memorable, while even some recent films have lapsed copyright and are thus available legally (but often against the film companies wishes) for videogram distribution. With the introduction of video discs by MCA in early 1979, this company's vast catalogue of movies has become available at only around $15·95, and many other catalogues will be becoming available during the coming years. On the other hand, there is currently a law suit being tried in the USA courts by MCA and Walt Disney against Sony for marketing a device (the video cassette recorder) that can be used to record movies off-air and thus cause damage to the producers of copyrighted programmes. Although in the short term this is obviously a problem (particularly with the American Home Box Office cable TV system that offers for a monthly charge first run movies at home), in the long term all these movies will become available on video disc at a price below that of the raw cassette – which then becomes uneconomic to pirate.

Pirating has also caused problems in Britain. It has been suggested that the day after a new movie is released in London, pirate video cassette copies become available on the black market. Presumably the projection copy is borrowed during the night and illegally copied! Copies of *Saturday Night Fever* were widely available at £500 and the film industry estimates that total worldwide losses account for a $700m loss. So obviously the sooner that such films and programmes become legally available as videograms on video disc, the sooner pirates will find the market uneconomic to satisfy on the more expensive video cassettes. The technology for video disc replication (pressing) is such that it will be feasible for only the best equipped companies to satisfactorily manufacture such discs (presently only MCA in California has the capability). The real change will have to be in the copyright laws otherwise many worthy programmes will never reach videograms.

6 Aerials

Roger Bunney

The aerial installation for a radio or television receiver comes to many as an expensive, if not unwelcome, addition to the overall system budget when new equipment is being planned. Too often economies are made with the use of an inadequate, poorly designed aerial, the result being inferior reception – noise, ghosts, distortion, varying signal levels on alternate channels – in short the high quality signal available from the broadcasters is not realised on the home receiving equipment. The most expensive equipment will be unable to provide good results, unless sufficient signal voltage is presented to the aerial input socket, and therefore wise policy will dictate the use of a well engineered and professionally erected aerial system, adequate for the receiving site in question.

Theory

A casual glance across the rooftops will show a vast number of aerial designs, shapes, sizes and varying degrees of complexity. It may be apparent to certain readers that the more elderly 405 line BBC aerials are large, the VHF-FM aerials smaller, the 405 line ITV aerials smaller still and the modern 625 line UHF aerials positively minuscule. To the more inquisitive it will be apparent that the actual lengths of the aerial rods vary according to the channel being received, being longer for the BBC 405 line channels (numbered B1–5 inclusive) with perhaps an overall dimension of 10ft, diminishing in size with increase in channel number (and increasing frequency) until the 'top' end of the 625 line UHF channels when rods of perhaps 6in are found (Fig. 6.1). Thus it is very apparent that aerial element dimensions directly relate to the frequency/wavelength it is designed to receive and thus for optimum results from a given transmitter, the aerial measurements

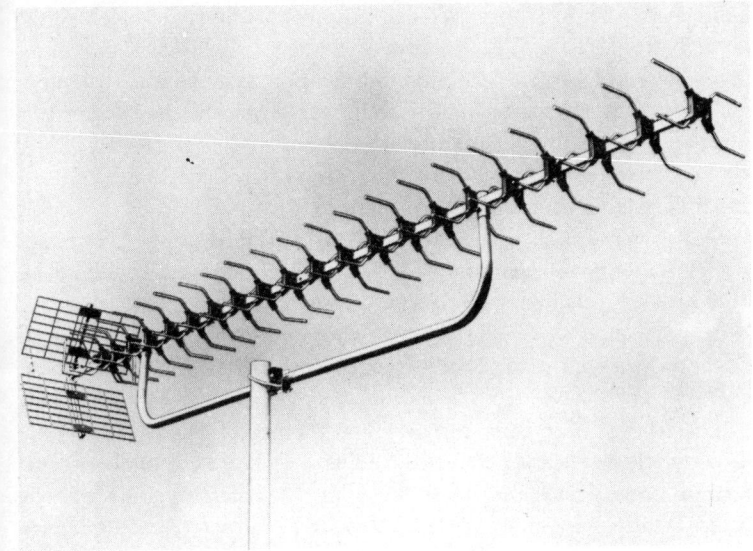

Fig. 6.1 Typical high gain group aerial from Jaybeam, with 21 multiple directors

must correspond to the frequency/wavelength transmitted. The signal energy as radiated from a transmitting aerial comprises two basic components, that of electric and magnetic fields. In diagrammatic form the fields are at right angles to each other, the electric field indicating the polarisation of the signal. The electromagnetic field radiates from the transmitter at virtually the speed of light (300,000,000 metres/s the actual wavelength of the transmission being determined by dividing the velocity (as above) by the frequency of that transmitter's oscillating energy (in Hertz). The Greek letter lambda (λ) is normally used for calculating and writing wavelength information.

If an aerial element is now placed within the electromagnetic field we will induce in this element sympathetic oscillating signal voltage/current information such induced oscillations being a maximum value if we ensure the length of the aerial matches the wavelength of the signal, or more simply the aerial is at resonance. It should be stressed that the element must be placed in the correct polarisation to obtain maximum voltage from the radiated electromagnetic field. For VHF and lower frequency applications, use of a full wave aerial could produce severe mechanical problems in supporting such an aerial structure. Fortunately a full current change can be accommodated in a half wave and normal practice will be use of a half wave length aerial at such frequencies. The half wave element is exploited for many domestic (and professional) communication systems and will normally take the form of a half wave element split in the centre, this separation providing terminations for connection of the feeder cable. The receiving element is normally termed a dipole and in the case of the half wave element is called a half wave dipole. From this point forward the half wave dipole will be simply termed 'dipole' since half wave elements will generally be discussed. In passing it should be noted that in practice the half wave dipole is slightly shorter than the 'free space' half wavelength due to a slowing of signal energy when flowing through a wire or rod element, the reduction in length being some 5%.

The feeder cable from the aerial to the receiver is usually a low impedance type such connection being made at the dipole central terminations where the impedance also is relatively low (approx 75Ω). It follows therefore that if cable with a characteristic impedance of 75Ω is connected to the dipole (also of 75Ω) a maximum transference of signal occurs due to the close match between cable and dipole. Providing this match can be maintained between dipole, cable and receiving apparatus, minimal transfer losses will result. Coaxial cable (unbalanced) is commonly used in the British Isles (Fig. 6.2a) but Continental practice often calls for 300Ω ribbon (balanced) feeder (Fig. 6.2b). Should imported equipment be used which features a 300Ω balanced input, a balance to unbalance (300/75) transformer (balum) must be used to maintain maximum signal transference.

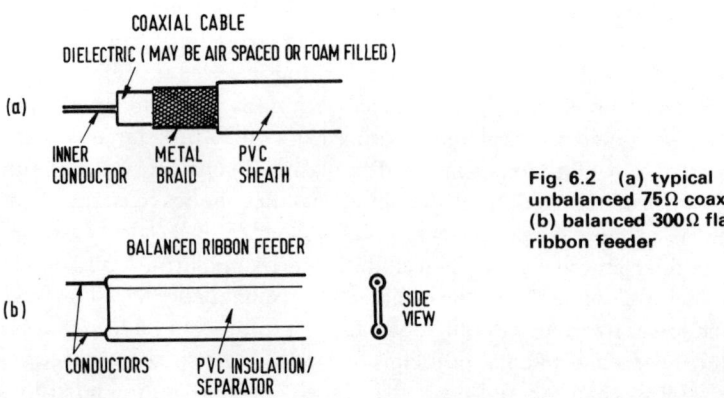

COAXIAL CABLE

DIELECTRIC (MAY BE AIR SPACED OR FOAM FILLED)

(a)

INNER CONDUCTOR METAL BRAID PVC SHEATH

BALANCED RIBBON FEEDER

(b)

CONDUCTORS PVC INSULATION/ SEPARATOR

SIDE VIEW

Fig. 6.2 (a) typical unbalanced 75Ω coax; (b) balanced 300Ω flat ribbon feeder

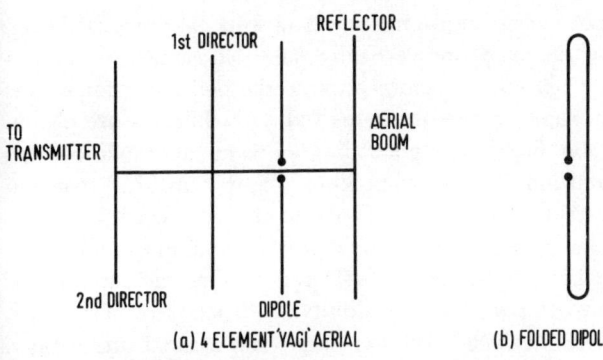

Fig. 6.3 Typical aerial detail indicating various elements

VHF and UHF signals from a transmitting mast generally provide good reception over a line of sight path which effectively means the distant horizon, as viewed from the transmitting aerial, is the limit of the service area. Generally the transmitter will be sited on high ground to maximise coverage, the main exception being relays that limit coverage to a specific town or populated area. In practice signals may be resolved just over the optical horizon by Tropospheric refraction, such signals being subject to fading and a degraded signal/noise performance. Receiving sites within the main coverage area that are screened by hills, tall buildings and other obstructions may also receive an inferior signal.

It follows that with increasing distance from the transmitter, signal field strength will fall. The dipole aerial, located in an area of high field strength near to the transmitter, will deliver adequate signal voltage to the receiver – but at a more distant site might be unable to induce sufficient signal voltage so means must be sought to increase signal output, or more specifically the power gain of the aerial system must be increased. If an additional (slightly longer) rod is placed behind the dipole at a calculated distance (commonly 0.15λ to 0.2λ) a signal improvement occurs due to the reflecting action of the new rod. The signal available at the dipole terminals will double (a power gain of 3dB). Still further power gains are possible if one or more rods, slightly smaller than the dipole, are placed at calculated spacings ahead of the dipole – these rods are known as directors, Fig. 6.3a. When such elements are added to the basic dipole structure various changes occur apart from the increase in power gain. If a vertically polarised dipole with omni-directional pickup characteristic (Fig. 6.4a) or polar diagram (for vertically polarised signals) is complimented with a reflector, a modification to the polar diagram occurs with signal pickup from the rear being reduced while forward pickup is increased (Fig. 6.4b). The polar diagram is further modified with director(s) addition, its forward pick-up lobe (beamwidth) sharpening with a considerable reduction in side and rear sensitivity (Fig. 6.4c). The reduced sensitivity to the rear is a very important characteristic in reducing co-channel interference. This measurement is determined by comparing the sensitivity of the forward pickup with the same signal received 'over the back', the aerial being rotated through 180° and measuring the drop in signal voltage. From this ratio is calculated the front/back ratio (in decibels). The director elements, as already mentioned sharpen the forward beamwidth and it is normal to express this figure in degrees, the limits being measured either side of the main forward lobe where the signal drops to half power (3dB) points. Very high forward gains are possible, it being common to use 18 or more director elements in a UHF array, Fig. 6.5.

Stress was placed on the degree of cable match to the dipole terminals. With the addition of a reflector, the dipole impedance will drop slightly (the closer the spacing, the greater the drop) but rather more dramatically when one or more directors are added. To maintain correct impedance matching into 75Ω coaxial cable, the dipole is usually folded (Fig. 6.3b) giving a four times increase in terminal impedance. In practice manufacturers adopt additional matching techniques since optimum matching must be maintained over several

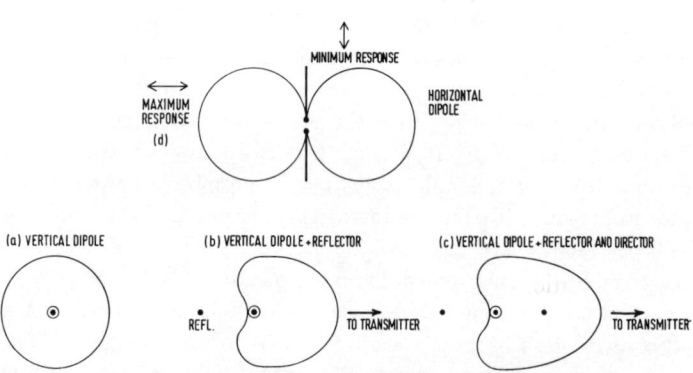

Fig. 6.4   Polar diagrams of simple aerials (viewed from above)

Fig. 6.5 Standard 18 element UHF Yagi aerial from Premier

channels. Unlike the 'early' 405 line VHF aerials that were designed to receive a single channel, modern VHF-FM and UHF aerials are designed to encompass a much wider bandwidth and considerable research and development is necessary in maintaining specified technical characteristics on factory made, mass produced systems.

For reasons of extra gain or for polar diagram tailoring (to remove ghosts etc.) similar arrays may be stacked and connected into one cable output. Such arrays can be stacked one above the other (vertical stacking) or side-by-side (broadside stacking). Depending on the spacing, power gain can be almost doubled (+ 2·5 to 3dB) with a reduction in either vertical or horizontal beamwidth. Up to this point vertical polarisation has been discussed, but in practice horizontal polarisation will be encountered for VHF-FM and UHF transmitters. The polar diagram of a horizontal dipole resembles a figure '8' (Fig. 6.4d) but the general rules of directivity are similar to that of vertical polarised arrays. Note: The Irish Republic's VHF-FM radio service uses vertical polarisation; most UK UHF relays operate vertically; BBC/IBA local radio stations use a variety of horizontal, slant and circular polarisation.

Aerial data

The half wave dipole is regarded as the base against which other aerials are compared and measured, and Fig. 6.6 indicates the charac-

Fig. 6.6 Characteristics of some typical aerials

Halfwave dipole aerial type	Power gain (forward)	Front/back ratio	Beamwidth (@ 3dB)
2 element, dipole+reflector	+3·5dB	8dB	±38°
3 element, dipole, reflector and director	+7·5dB	30dB	±32°
4 element, dipole, reflector, 2 directors	+9dB	20dB	±20°
10 element, dipole, grid reflector, 8 directors	+11·7dB	28·3dB	±21°
18 element, dipole, grid reflector, 16 directors	+14·7dB	30·7dB	±16°

teristics of some typical aerials. It will be noted that as a general rule, doubling the size of the aerial will double the gain (+ 3dB), clearly a law of diminishing return since mechanical considerations will limit the eventual size (and hence gain) of an aerial structure.

Reference levels

With recent advances in electronic technology, receivers are now capable of working with signals of relatively low field strengths and giving acceptable quality results to the listener or viewer. Unfortunately with lower levels of signal, interference sources (electric motors, ignition etc.) become more noticeable and it is essential therefore to provide the maximum input signal level to mask any possible interference breakthrough. In order to achieve an optimum signal to noise figure with consideration to both man made noise (interference) and internal receiver noise, certain minimum figures have been suggested as receiver input levels for both VHF-FM radio and for UHF television. At VHF 500µV (0·5mV) should ensure a good quality stereo signal within the 88-100MHz spectrum. This relatively high figure ensures an optimum signal to noise performance for stereo, which by virtue of the additional information carried in the stereo channel will need a somewhat higher signal level to prevent an intrusive 'hiss'.

The demands of colour and teletext decoding are such that higher input signal levels are necessary than for monochrome television. The professional aerial engineer will design a system to give 1000µV (1mV) in Band 4 and 1500µV (1·5mV) in Band 5. The low noise high gain UHF tuners now available are capable of resolving a quality performance at lower input levels, adequate monochrome reception being possible at 300µV (0·3mV) input. Fig. 6.7 show which aerial groups come within which bands of channels.

Fig. 6.7   Aerial groups and channel bands

Band 4		Band 5	
Group A Channels 21–34	Group B Channels 39–53	Group C/D Channels 48–68	Group E Channels 39–68

Amplifiers

In many locations the available signal strength will be too low to provide adequate quality reception and amplification is necessary to lift
the weak signal to an acceptable level. Aerial amplifiers are commonly used in fringe areas to improve signal strength and the signal
to noise performance of the receiving chain. The amplifier (sometimes known as preamplifier) is often fitted externally and adjacent to
the dipole prior to the loss that arises in the coaxial downfeeder.
Amplifiers are available as either 'grouped' (that is designed for single
channel Groups A, B, or C/D) or wideband (covering the complete
UHF Spectrum), and with voltage gain figures of 12dB in a single
stage 'group' unit, to perhaps 22dB with a two stage wideband unit.
Labgear is now producing two stage 'group' amplifiers for aerial
mounting with a gain of 29dB in Group A (Fig. 6.8). Apart from
quoted gains, other specifications of importance relate to the noise
figure and the amplifier's cross modulation performance. All amplifiers unfortunately contribute noise to signal information passing
through, resulting in 'hiss' on sound or a light drizzle or snow affect
on vision. If a signal input is marginal and somewhat noisy then it is
of utmost importance to use an amplifier of very low noise to improve
the overall system signal to noise performance, particularly if the
receiver is of an early type with a relatively high noise figure. Such a
receiver will benefit considerably from a low noise preamplifier while
the latest 'state-of-the-art' receiver with a low noise front end, may
show a minimal improvement.

Cross modulation is another problem that can occur with an amplifier if presented with too high a signal input level, the effect being
to give (on vision) excessive contrast, the local UHF channels 'floating' simultaneously on top of each other; a spreading of the local
signals onto adjacent vacant frequencies and on sound a loud video
buzzing (varying with picture content). Manufacturers usually quote
maximum signal input levels above which cross modulation will occur.
Figures are often quoted relative to 1mV and an understanding of the
decibel and its gain ratio figures is helpful. An amplifier gain 6dB

93

Fig. 6.8 Labgear mast mounting UHF amplifier with 27dB gain

voltage is twice, 12dB is 3·98-say four times, 18dB is 7·94-say eight times, and so on. With 1mV used as reference, these figures convert to 2mV = 6dBmV; 4mV = 12dBmV; 8mV = 18dBmV.

To further expand this important subject, the Labgear Technical Applications Brochure is enlightening:

Type CM7025 Group A (two stage), bandwidth Group A 470-581MHz, voltage gain 29dB, noise figure less than 4dB, maximum handling capability for 1 channel 63mV (36dBmV) or 4 channels 35mV (31dBmV) and input/output impedance 75Ω.

Type CM6000 Group A (one stage), voltage gain 16dB, noise figure 3·5dB, maximum handling capability for 1 channel 25mV (28dBmV), 4 channels 14mV (23dBmV), and input/output impedance 75Ω.

Reception difficulties

Despite the large number of main and relay transmitters, all manner of aerials, amplifiers and high performance receivers, a sizeable number of the population suffer from inferior reception. Such degraded reception can arise through weak signals (fringe or screened locations), man-made interference, multiple images (ghosts) and interference on the same and adjacent channels, from other distant transmitters.

It is apparent from the many letters seeking advice received by the writer, that the number of 'professional' aerial engineers active within the domestic sphere are minimal. Most parts of the United Kingdom have adequate field strength from one or more transmitters and a good signal can be resolved with a basic array which requires little technical skill or assessment on the part of the rigger. Thus has arisen in the trade the term 'cowboy rigger', happy at the location which presents no difficulty but less happy when presented with a problem. Often his charges are relatively low, resulting from cheap equipment, a rapid installation, the criteria being quantity rather than quality. An example recently was an installation the writer was called in to repair: a 6ft steel mast supporting 18 and 12 element UHF arrays, fixed against a wall with a standard bracket – the latter held with three 1in wood screws and plastic wall plugs into soft mortar!

The reader is advised, when seeking an aerial engineer, to establish if he uses a recognised 'professional' measuring instrument such as the TES field strength meter, and to insist on certain measured readings at the aerial output socket(s) of: VHF-FM 0·5mV on each channel; UHF Group A 1mV; Group B, C and E 1·5mV. The experienced engineer will appreciate the customer's attention to such standards and can be relied on to give a satisfactory installation. The problem of weak signals commonly result from either sheer distance from the transmitter or from local screening (the receiving site is behind a hill). When ghosting is not present, the basic quest is that of initially optimising aerial power again and supplementing signal output with a low noise preamplifier for the frequency spectrum concerned.

In the early days of UHF reception, the highest gain array was perhaps the single in-line yagi with approximately 20 straight director elements. Extra gain was then obtained by stacking a similar array and combining signal output with phasing bars or a cable harness and achieving a 2·5dB to 3dB power gain. Considerable research into aerial technology is an ongoing activity in the industry – always seeking improved gain and overall performance for a given amount of metal structure and other material. In West Germany, during the late 1960s, there a system was evolved using full wave multiple directors, the theory being to achieve the gain of a stacked array but with only one aerial support boom. This system has been adopted universally with individual manufacturers applying their own variations on a basic theme. With the potential of really high aerial gain available along a single support boom, manufacturers have introduced relatively large arrays featuring up to 21 multiple director assemblies, one manufacturer has available a phasing harness to stack four such multiple director systems! The approach to weak signal reception at the distant (fringe) location is that of basically high gain aerial systems and as has been shown, such performance is available from many companies. The addition of a low noise aerial amplifier, particularly if a long cable run is anticipated, will prove beneficial and certain amplifiers (such as the Wolsey UHF Supa Nova) are noted for their performance under weak signal conditions. Certainly 'the bigger the better' for fringe work is the rule to obtain an acceptable signal to noise performance.

Stacked aerials, whilst improving aerial forward gain, will also benefit with interference reduction by virtue of the sharpened forward lobe and high attenuation of unwanted signals from the side. In weak signal work, the co-channel or adjacent channel interference signal is one of the major problems and where budgets allow the horizontal stack should be advocated. Perhaps the advantage with a fringe area are the possibilities of at least one alternative IBA station, in such cases a fixed high gain array for the alternative should be used, fed with separate low loss feeder, selection indoors via a low loss switch for the appropriate channel. With a multichannel installation head amplification is an ideal for both aerials but home economics may dictate otherwise. The alternative is to use very high quality coaxial feeder to minimise losses and fit the aerial amplifier after the selection switch, the wideband UHF amplifier being advocated rather than the wideband VHF/UHF amplifier, again in the interests of minimising possible radio interference breakthrough (Fig. 6.9).

It is often possible, given a favourable high location, to provide a second IBA channel well within the service area of the local IBA station. Whereas high gain aerials again are essential, care must be taken particularly if the distant signal is within the same aerial group as the local. The aerial system must possess excellent rejection in the direction of the local transmitter, often a horizontal stacked system being necessary if the local signal arrives on the side of the aerial structure. The head amplifier must be of a type that has a good cross modulation and overload performance since a distant signal may be sought on an adjacent channel to the local channel. There are UHF notch filters available that give a high degree of attenuation on a given channel with only a nominal loss on its adjacent channel. If several alternative channels are available from various directions, it may prove more economic to use a wideband multiple director system array, a low noise high gain (with low cross modulation properties) amplifier together with an aerial rotor system. The aerial rotor is an electric mast mounted motor fed with multiway cable from an indoor control box and capable of rotating an aerial through 360° to a required direction. Various types of aerial rotor are available covering a range of prices (prices generally relating to headload) but care must

Fig. 6.9 Labgear indoor distribution amplifier for two sets

be taken in both selection of the type to use and its mounting on top of the mast, always erring on the side of safety.

The weak signal at a screened location tends to be somewhat more difficult to resolve than the fringe signal. Consider Fig. 6.10 where a house is severely screened and receiving an indifferent reception quality, mainly from diffraction across the crest of the hill. It follows that by increasing the height of the array into the diffracted signal field, the aerial output voltage will correspondingly increase and that effort must be placed into maximising aerial height. It is very likely however that the signal quality over the channels within the required group will fluctuate considerably, and with difficulty in equalising signal voltage over these channels. It again follows that if two aerials are stacked one above the other, the problem will ease considerably since signal pick-up will occur over a wider capture area and the output of the system will reflect a much more level response. One aerial

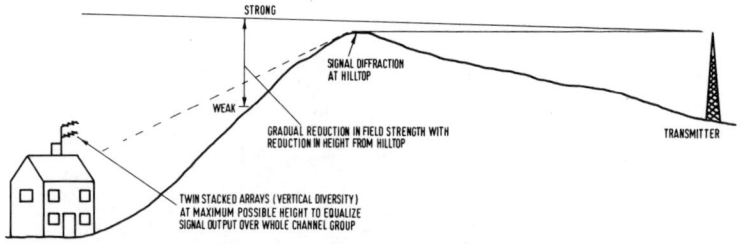

Fig. 6.10   Reception difficulty in screened location

that has found application with this type of work is the stacked bowtie, a wideband aerial of simple design but excellent performance.

Ghosts

A signal reflection or ghost is a common occurrence in built-up and hilly areas whereby a second (or more) reflected signal is received shortly after the direct primary signal. Since the secondary signal arrives fractionally later than the primary, it is visible on the receiver screen as another usually weaker picture and slightly to the right of the main image, the distance between the two images relating to the additional distance the secondary signal has travelled via the reflecting object. Since the ghost is arriving on the aerial response pattern by reflection, it follows that the secondary signal is arriving from a direction other than the primary signal. If a highly directional yagi system (usually upwards of 12 elements at UHF) is employed, it is usually possible to minimise such pickup. The polar diagram or plot of a typical yagi is not smooth but displays side lobes and nulls. The aerial can therefore be rotated to place the ghost in a null, provided adequate forward pickup from the primary signal can be maintained.

The log-periodic aerial (Fig. 6.11) is another type that has all its elements 'active', is wideband and features a very smooth polar diagram, free from secondary lobes. It can be employed successfully in anti-ghosting work but its gain, unfortunately, is low. As mentioned earlier the conventional yagi can be stacked to provide not only improved forward gain but a polar diagram with greatly reduced side lobes. Such a horizontal stack may have the spacing of the two arrays adjusted to 'steer' nulls into specific directions where ghost signals are known to be originating, thus minimising their pickup relative to the main forward lobe. The aerial manufacturer should be consulted for advice detailing the type of arrays in use and the direction that it is

98

required to reject. He will have the detailed polar plot and can advise a suggested spacing for the two aerials. The harness for combining the two arrays is similar to that used for stacking as described earlier, under normal circumstances sufficient cable will be incorporated in such harness to give a wide variation of spacing for locating the correct phase cancellation and the loss of the ghost. Teletext causes particular problems (**Fig. 6.12**).

Reflected signals that are somewhat more difficult to eradicate concern those locations adjacent to busy airfields when signal reflection from moving aircraft can occur. By virtue of the ever-changing position of an aircraft in both height and direction, only a compromise attempt can be made to reduce such a problem. The vertical stacking of two similar arrays of at least 12 elements and spaced by at least two wavelengths will help to restrict vertical beamwidth and to a degree minimise such continually fluctuating picture distortion. The rear mounting aerial will facilitate mounting on the conventional vertical stub mast. The above general guide lines also relate to VHF-FM radio reception, but certain of the mechanical structures suggested will be impossible to achieve due to the considerably greater dimensions of the standard Band 2 aerial (**Fig. 6.13**), the elements being some 5ft in length. Multipath reception on FM radio manifests itself as distortion, its severity relating to the strength and phase relationship between the direct and reflected signals.

Fig. 6.11   Wideband log periodic aerial (Premier)

Signal distribution

With the increasing number of 'two set houses', video cassette recorders and VHF stereo radio, the need for two or more signal outlets – probably in different rooms – is growing. If two outlets are required for television, and given an input of perhaps 2mV at UHF, a simple 'star' splitter can be made to feed separate outlets (Fig. 6.14). The simple 'star' network will give a reasonable match into two separate 75Ω outlets; a fundamental rule with this and all distribution systems is that accurate matching must be maintained at 75Ω throughout the system. On no account should direct taps be made into an aerial feeder since this can give rise to mismatch, line reflections and signal loss, particularly if the outlets are left unterminated (unconnected).

It follows that if an input signal is split two ways, each of the two outlets receives a maximum of half the input signal – ignoring splitter and cable losses – in the case of the 2mV input signal an outlet voltage of 1mV would be present (−6dB loss). There is every possibility in the modern house that perhaps four TV outlets are desirable, with a further two VHF radio outlets and this would call for a high input

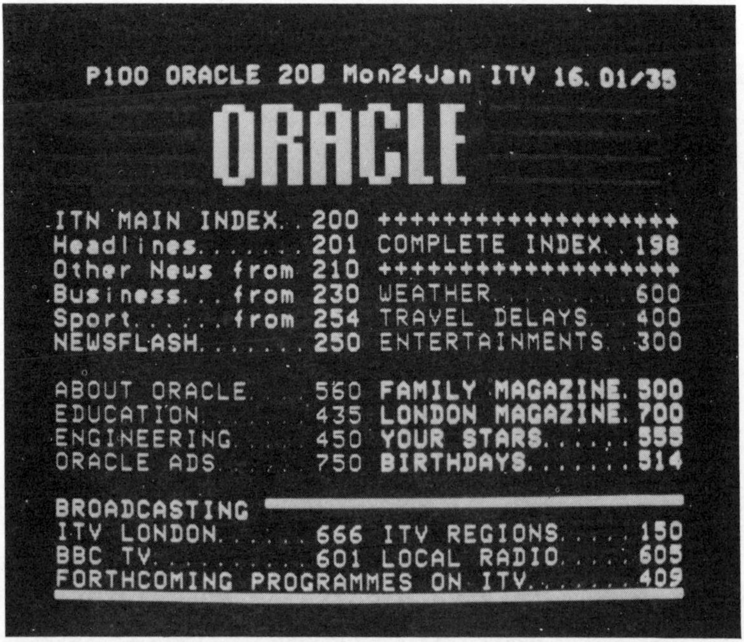

Fig. 6.12a Teletext page correctly displayed

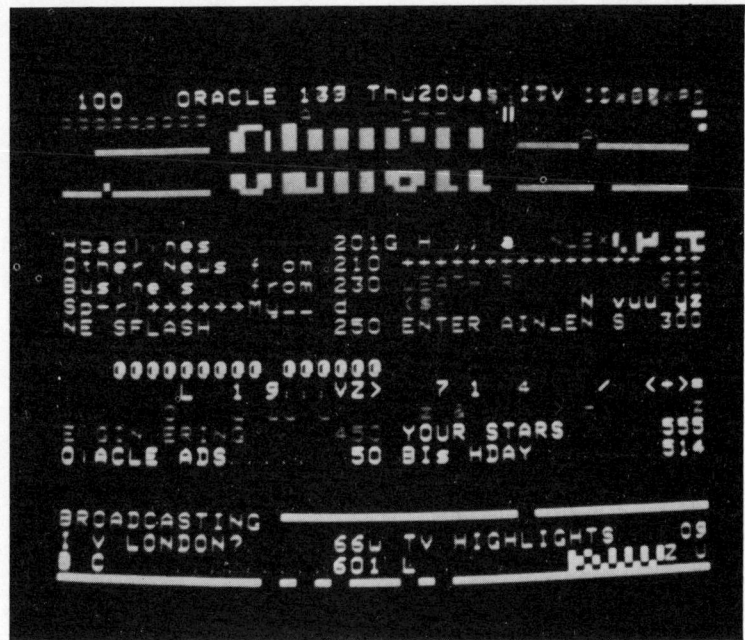

Fig. 6.12b Teletext page where ghosting causes lines to be missed and characters misdisplayed

signal voltage from the aerial array. Fortunately several aerial manufacturers are now producing domestic distribution amplifiers that solves the problem of perhaps six outlets in a single dwelling. In such a system the single or composite input signal is amplified in a wideband circuit (40–860MHz), thus embracing the Bands 1 and 3 405 line channels, Bands 4 and 5 625 line channels and Band 2 VHF radio. Typical amplifiers have upwards of four outputs at perhaps + 6dB relative to input, the gain helpful in overcoming feeder losses within the dwelling. At each signal outlet, there will be present an adequate voltage level reflecting the composite signal input and sufficient to feed modern domestic receiving equipment provided the input signal to the system is within the guidelines as detailed earlier.

Should a television receiver be used in conjunction with a video cassette recorder, it is normal practice to feed the input signal into the recorder, internal amplification being provided to feed both the internal tuner and an output socket for connection to the television receiver, the output socket carrying the input signal at unity or a low

Fig. 6.13 Typical high gain (9·5dB) Band 2 six-element aerial (Antiference)

level gain. Figs. 6.15 and 6.16 show two different types of domestic splitting situations.

Aerial hardware and components

The casual glance across the rooftops mentioned earlier across the landscape of dipoles, directors and reflectors will show a variety of masts, chimney lashings and the like – hardware to retain the aerial 'in situ' throughout all types of weather and over a considerable number of years. Clearly good quality components must be used, masts rated to carry their headload in winds of perhaps storm force, lashings that will retain their position and without movement of the

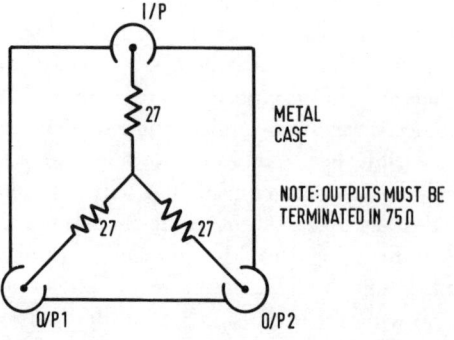

I/P

27

METAL
CASE

NOTE: OUTPUTS MUST BE
TERMINATED IN 75 Ω

27   27

O/P1   O/P2

Fig. 6.14 Simple star
network splitter (or
combiner) for VHF use

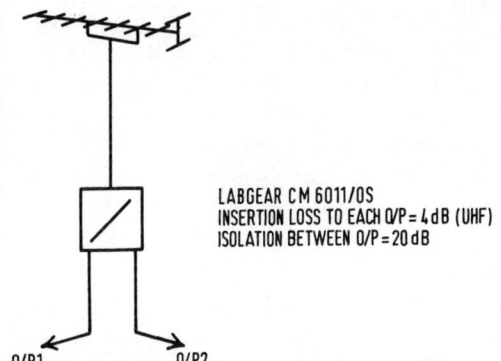

LABGEAR CM 6011/0S
INSERTION LOSS TO EACH O/P = 4 dB (UHF)
ISOLATION BETWEEN O/P = 20 dB

Fig. 6.15 Ferrite core wideband splitter (or combiner) for VHF/UHF

O/P1   O/P2

lashing wire around the chimney, and wall brackets secured by bolts into a hard brick. Coaxial cable carrying the signal from the aerial to the internal socket must be high quality low loss feeder since a considerable amount of signal can be lost through incorrect use of poor cable, with sharp bends and compression under too deeply hammered staples all contributing to loss. Clearly the professional aerial engineer will carry out installation to a high standard, again must be emphasised 'professional'. Other ancillary components that must be used in ensuring efficient signal transference are the $300\Omega/75\Omega$ matching transformer for VHF-FM receiver use when imported equipment is

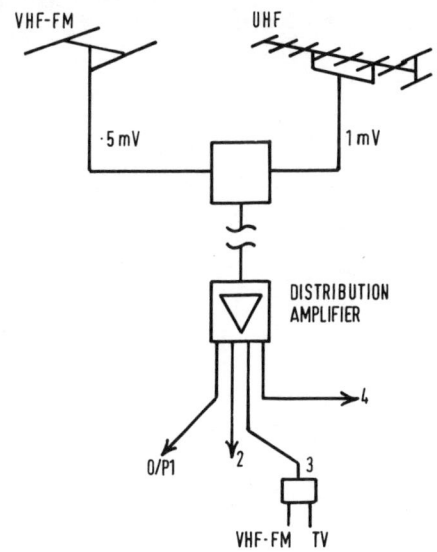

Fig. 6.16 Comprehensive domestic installation combining two aerials, and using a distribution amplifier to feed four rooms, with one room also having a splitter to feed radio and TV

in use and such transformers are available from most aerial manufacturers. The advent of the television game often results in frequent replugging at the receiver aerial socket rapidly causing wear and intermittent contact and several companies have introduced a wideband ferrite coupling transformer enabling two inputs to be fed into the single receiver input socket, the transformer providing a high degree of isolation between the two inputs.

Manufacturers standards

The television and radio aerial is the first tuned circuit in the receiving chain and clearly an efficient aerial will ensure optimum signal pick-up for processing in the receiver circuits, hopefully resulting in excellent programme quality as the end product. The aerial manufacturers being aware that a minimum standard should be established for the manufacture and assessment of aerial equipment, formed an association to decide on minimum standards and methods of testing and measurement. The British Aerial Standards Council was formed, members of BASC maintaining test facilities and staff on each of their premises for on-going testing of samples from the production line to retain prescribed standards and for development work.

At the time of writing, BASC members are Aerialite Ltd, Antiference Ltd, Jaybeam Aerials Ltd, Maxview Aerials Ltd and Wolsey Electronics. A number of smaller manufacturers that do not maintain staff for test and evaluation formed themselves into the Aerial Manufacturers Association. The AMA does however lay down standards for aerial design, construction and performance, together with minimum standards for associated hardware such as brackets, clamps etc. Aerials are tested and evaluated by the University at Manchester, an independent body; membership of the AMA being dependent on certain laid down technical standards. Premier Industries (Cheltenham) Ltd is currently spokesman for the AMA.

Information

The BBC and IBA both have transmitter lists, reception notes and other information available free of charge on application to the following information sections:

Engineering Information Department, BBC, Broadcasting House, London W1A 1AA.

Engineering Information Department, IBA, Crawley Court, Winchester, Hants, SO21 2QA.

Conclusion

It is hoped that the information in this chapter will have helped to enlighten the reader into the requirements and standards for an efficient receiving installation. The efficiency, mechanical quality and installation standard are all of equal importance since wise invest-ment at this early stage will result in a system that is capable of giving years of service with a maintained performance throughout its useful life.

7 Television Receivers

Angus Robertson

Whilst there has been no remarkable improvement in television receivers since about 1970, it is certainly true that each year brings about
an evolvement of many aspects of receivers, with the result that the
colour receiver of today provides an extremely stable picture of high
quality and excellent reliability – and at little more cost than 10 years
ago. Many new features are becoming available, such as remote
control (Fig. 7.1) and automatic tuning. Television receivers can now
be purchased complete with teletext and viewdata decoders, and even

Fig. 7.1 Typical modern colour television from ITT, offering remote control

with built-in television games. Fortunately, technology has been advancing at such a rate that all the additional circuitry required to provide new facilities can be added at little extra cost. In fact, inflation means that in 1979, colour televisions are now about a third of the price in 1969.

Basically, a television set comprises two separate systems – a television receiver and a television display. These elements can be considered to be two separate parts, as in the case of hi-fi where one has an FM receiver/amplifier and separate loudspeakers. Since televisions still have primarily the single role of viewing programmes transmitted by broadcasters, all consumer television receivers are currently self-contained with receiver and display in a single cabinet. In education and industry, television monitors are widely used and these contain only the display circuitry which may be fed at 'video level' from a video tape recorder, distribution system, camera or separate television receiver unit. Receiver/monitors (Fig. 7.2) are also available and these offer the best of both worlds – a television receiver that also offers video inputs and outputs.

Fig. 7.3 is a schematic layout of a typical colour television receiver. This shows the basics and can be variously modified depending upon the type of tube and various features provided. The television aerial is connected to the tuner module which takes the ultra high frequency (UHF) transmitted television signal, amplifies it and modulates it down to a lower intermediate frequency in the 30MHz region. The tuner modules available in the early days used mechanical tuning with a variable capacitor driven from a rotary tuning control, but the introduction of varicap diodes (whose capacitance may be varied by altering the voltage across them) permitted mechanical tuning components to be eliminated and tuning was then accomplished by switching voltages (typically up to 33V). This voltage is derived from the channel selection module. In most modern sets this contains either a number of mechanically selected variable resistors, which provide variable tuning voltage depending upon the channel selected (usually by depressing and twisting the appropriate tuning button), or 'touch tuning' which still requires a bank of variable resistors (this time electronically controlled by integrated circuits activated by two touch contacts shorted together by body resistance). Variable resistors (potentiometers) are still mechanical and often cause problems in receivers, so the latest technique being used widely in Europe uses totally electronic control, and is discussed with remote control later.

The intermediate amplifier module provides more gain and selectivity to discriminate against adjacent signals and interference, the IF (intermediate frequency) signal then being passed to the video detector, which provides baseband video separated into luminance (the brightness of the picture) and chrominance (the colour). At this point, the video signal is also taken to the audio IF and detector where the sound channel is detected and passed to the audio amplifier. Full details of television sound are contained in Chapter 2, but here the one point that should be made is that volume controls have also been changing. Although most volume controls in radio receivers take audio actually to the volume control, recent television receiver volume controls provide a DC voltage (usually the higher, the quieter), which is applied to the audio IF and detector module to vary volume level. This has two advantages: only a single unscreened wire from front panel to circuit board rather than two screened cables; and the volume can be easily remote controlled.

Since the transmitted television signal must be compatible with both monochrome and colour viewers, the colour information (chrominance) is transmitted separately and can be seen entering the

Fig. 7.2 Barco colour monitor/receiver

108

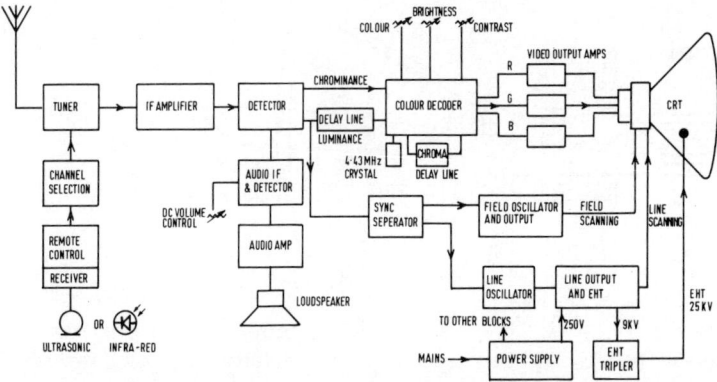

Fig. 7.3 Schematic typical colour television receiver

colour decoder block together with luminance (brightness), which is delayed to compensate for all the colour processing. This effectively delays the colour signal by an equivalent amount. The colour decoder separates the luminance and chrominance into red, green and blue signals which are amplified to drive the cathode ray tube display. The DC contrast, colour and brightness controls also operate within the colour block, and other items include a 4·43MHz crystal for decoding the chrominance and another delay line which delays colour from one television line so it may be compared for errors with that transmitted on the next line.

Having dealt with the signal processing circuitry of a typical television receiver, the scanning circuitry necessary to display the signal is examined. Chapter 2 described the principles of television scanning which comprise line scans taking the beam across the screen and field scanning taking it up and down the screen. These scanning drives must be locked to the television signal and this is achieved using the sychronising pulses transmitted with the picture (Chapter 2 again) which are removed from the television picture by the 'sync separator' and used to lock both the field and line oscillators, the outputs of which are amplified to drive magnetic coils around the neck of the cathode ray tube (CRT) to provide scanning. While field scanning is a conventional amplifier, the line output amplifier is also used to generate extra high tension voltage required by the CRT to provide picture brightness. The power supply normally produces around 250V, which

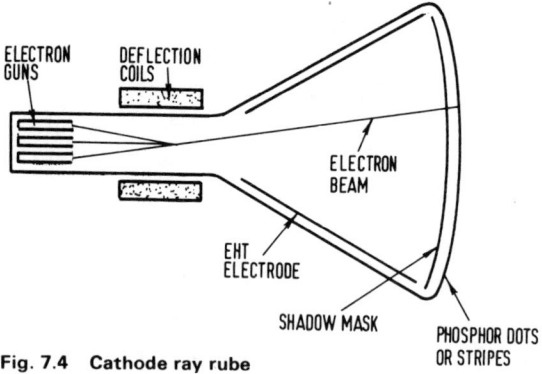

Fig. 7.4 Cathode ray rube

is adequate to drive most television circuits. Thus a transformer is required to boost this to 9kV (typically) from where a special 'tripler' circuit is used to generate about 25kV. A transformer operating at the mains frequency of 50Hz would be enormous and heavy, so the line output transformer (which is operating at a frequency of 15,625Hz) has an extra winding which can thus be very compact (due to the high frequency involved) and this produces the 9kV.

Although the CRT has been mentioned, it has not been described. Fig. 7.4 shows the principle of the colour CRT. There are three separate electron guns for each of the colours, red, green and blue. The guns generate a beam of electrons whose intensity is varied by the video output amps shown in Fig. 7.3. After focusing, the deflection coils, which are driven by the field and line scanning drives, deflect the three electron beams across the screen. In a monochrome tube, there would be only a single electron gun, which would be scanned so as to illuminate the phosphor coating on the inside of the CRT face-plate, which in turn produces light and a television picture. With colour tubes, there are three different coloured phosphors on the tube, which when illuminated to varying intensities, can produce all the colour necessary to produce 'colour television'. So as well as three electron guns, there must be three different coloured phosphors on the tube front, and these must be arranged so that electrons from one gun, only 'hit' the coloured phosphor they are intended to illuminate, otherwise colour errors rapidly become apparent. The first colour tube was developed in 1950 by RCA and was called the 'shadow mask'. This used groups of red, green and blue phosphor dots etched onto the screen, which were only illuminated by the respective electron beams; a shadow mask with holes prevented each beam hitting other dots (Fig. 7.5a). Typically there might be almost 500,000 holes in the shadow mask and 1,500,000 dots on the screen – but when viewed from normal distances these all merge into a colour television picture. In the shadow mask tube, the three electron guns were mounted in triangle formation to correspond with the dots, and this meant that considerable electronic and magnetic correction was required to produce a correctly converged picture. In other words, the three different coloured pictures had to overlap each other precisely, otherwise 'fringing' (a coloured edge) was seen, often on white objects. In such a system, some 14 controls were required to align a shadow mask tube and these all tended to drift with time, and were tedious and time consuming to set-up.

Not wishing merely to be a licensee of RCA, Sony spent many years developing television tubes. Initially they came up with the Chromatron in 1964 (under patent from Paramount) which used a single gun electronically switched as it swept past each different coloured phosphor. Of course this was a considerable simplification on the shadow mask, but had a variety of other problems which made Sony abandon it. Eventually, in 1968, they introduced the Trinitron tube. This used three guns in-line (an idea initially proposed by General Electric), but rather than using dots, the Trinitron uses vertical stripes of phosphor on the screen, with an aperture grill of vertical strips (Fig. 7.5b). This

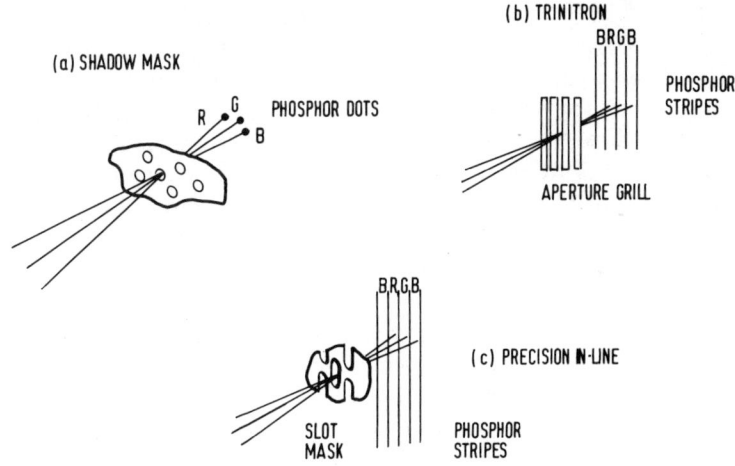

Fig. 7.5   Three different colour CRT techniques

had a great advantage over the shadow mask which was very inefficient since many of the electrons merely fell on the mask, and did not travel through holes. However, although Trinitron pictures were very bright with few convergence problems, it was not possible initially to manufacture tubes larger than 13 inch because the vertical strips moved, and so the shadow mask tube remained supreme in the outside of Japan where 20 inch and larger screens made up 90% of colour sales.

Although 20 inch Trinitrons are now available, another development, the Precision In-Line (PIL) tube was introduced in 1975 and this again uses three electron guns in-line and vertical phosphor stripes, but this time the mask contains slots which make it rather more robust than the Trinitron and thus available in sizes up to 26 inch (Fig. 7.5c). The PIL tube construction is such that focus is well maintained over the entire screen, and the tube is self-converging with no electrical convergence controls, as these are merely preset during manufacture. This eases manufacture and servicing considerably, and allows sets to be manufactured more economically.

So what new developments are there in colour television receivers? Remote control is the most obvious addition to recent television receivers, and is fast becoming a necessary selling feature. Basically there are two types of remote control, each implemented by one of two methods – ultrasonic or infra-red. Ultrasonic remote controls were the earliest development and utilise sound of frequencies above that which the human ear can hear, at around 40kHz. The simplest units merely transmit one frequency for say channel change on a cycling basis, i.e. one channel after another. Many sets use two frequencies allowing the television sound to be cut or muted from the remote control (very useful for killing commercials and repetitive stations identifications). There are some ultrasonic remote controls that allow up to 30 different functions to be controlled, typically selection of 16 channels, volume, brightness and colour increase or decrease, sound mute, and television power on/off (Fig. 7.6). In these multichannel units, 30 different frequencies are generated from a crystal and each can be determined in the receiver.

Unfortunately, ultrasonic remote control has one drawback; animals like cats and dogs can hear the high frequencies used, which have been known (very occasionally) to cause 'nervous breakdowns' in highly strung pets. So the latest remote control units use infra-red light, which the human eye cannot see. It is specially modulated with

Fig. 7.6 Ultrasonic remote control with is normally stowed in set front-charging batteries

the required information in order to provide similar facilities to that of ultrasonic sound. In my experience, the earlier infra-red units were not very satisfactory since the remote control unit had to be pointed directly at the receiver so as to gain a response. In contrast, ultrasonic sound bounces easily off walls and furniture (but not off cushions or curtains) and is thus far less directional.

One reason for remote controls being required is that manufacturers still insist on manufacturing televisions as self-contained units. I expect to see televisions available with separate receivers and displays within a few years, with the increase in video cassette recorders, games, teletext and viewdata, and video discs when plugged into an video socket. They all provide higher quality than if plugged into the aerial socket. Remote controls have become an economic possibility with the introduction of specialised integrated circuits (chips) containing much of the complicated circuitry. Although users in Britain rarely require more than six channel pre-selectors on televisions (BBC1, BBC2, ITV, another ITV region, video cassette, and a TV game), in other European countries many other channels are available. For instance many parts of Belgium have television available from the Netherlands, France, Luxembourg, West Germany and even Britain. One can see why 16 preselectors are no longer sufficient for many sets; 20 is becoming normal. However, this number of preselectors means that 20 variable resistors are required for channel selection, and numerous switching circuitry to operate it all.

So the trend is now towards totally electronic tuning, where the information required to produce the tuning voltage is stored in electronic memory, and recalled when required. Automatic tuning preselection is usually supplied where one merely presses a button, and the television tunes through the band itself, stopping at the first receivable channel it finds. If this is not the one required, the button is pressed again, and the next channel appears, and so on. This is repeated for each preselector until all the different channels required have been found. Other buttons allow a preselection on one channel to be transferred to another button, or for fine tuning adjustment. Automatic frequency control ensures that channels remain correctly tuned. However, with the introduction of up to 20 preselectors, then 20 buttons each with a light are still required and this becomes complicated electronically. It might also be considered that when many channels have been preselected, one might just as easily select the actual channel number required. Thus the latest sets simply have a 0 to 9 digit keypad upon which the required channel number is entered, the selected number then appearing in the corner of the TV screen for a few seconds after selection. Alternatively, a tuning scale might be displayed with a horizontal bar indicating channel. It will even be possible to press a button on the remote control which will display on the screen a list of channels actually available with their service name, from which the selection may be made.

In conjunction with teletext, it might eventually be possible to display on the screen a list of programmes actually being transmitted at any moment in time, onto which the receiver will add channel numbers (since the same information might be transmitted via many different relay transmitters operating on different frequencies). In the USA, some televisions are available which may be pre-programmed with the times of specific programmes during the week, and which will then only operate the television during those periods – ideal to stop the kids watching just anything!

It is unlikely that the conventional television will become available with any larger than a 26 inch (diagonal) screen, due to the problems of manufacturing larger glass envelopes for the CRT. Sony has constructed a prototype 32 inch receiver, but it is unlikely to be marketed. Any developments in large screen television are likely to be in the area of projection television (and are covered in Chapter 9). Picture-within-a-picture is another feature demonstrated, but not as yet marketed. It allows a second channel or video input to be display-inserted as a small picture in one corner of the main picture (Fig. 7.7). Although it might sound simple, in practice it requires that the inserted picture be electronically stored because it will not be in synchronism (i.e. the same lines arriving at precisely the same microsecond as those of the main picture) and thus would not appear correctly placed. However, technology now enables this to be performed economically, and it will probably make an appearance in many sets over the coming years. Meanwhile, Nordmende have a receiver which incorporates a 26 inch colour set, and three 8 inch monochrome screens (Fig. 7.8) which may be separately tuned to different channels! Another feature is infra-red cordless headphones that allow the television sound to be picked-up by headphones with an infra-red receiver built into the headband.

Fig. 7.7   Picture-within-a-picture; a Barco development

Fig. 7.8   Nordmende colour receiver with three monochrome
screens

Chapter 1 indicated how the number of colour television receivers
is rapidly growing in the UK, while the number of monochrome is
severely falling. Many of those that do exist are second sets, mainly
imported from the Far East. But small colour sets are also making an
impact as second sets, and typical of these are a British built ITT 16
inch set (Fig. 7.9). One of the biggest problems with all the interest in
television games, teletext and video is that the main set becomes over-
worked, and can only offer one alternative at a time. The second set
obviously relieves this problem.

Finally, it might be as well to examine briefly the state of television
broadcasting in Britain in order to understand why many old sets are
unreliable or becoming obsolete. When high quality television first
started, it was broadcast by the BBC on VHF Band 1 (Fig. 7.10), using
405 lines. When ITV came along, Band 3 was opened to accommodate

Fig. 7.9 British-made portable 16-inch colour television

the extra channel, but also on 405 lines. During the 1960s, it became necessary to change to 625 lines for colour transmissions, so a totally new transmitter network was built based on UHF in Bands 4 and 5. BBC2 colour opened in 1966, but only on Band4/5, while BBC1 and ITV remained on Bands 1 and 3 until 1969; then colour transmissions on Bands 4/5 were added in duplicate to the existing VHF bands. So television receivers manufactured before 1965 operated only on 405 lines VHF, while those built between 1965 and 1969 were dual standard operating on both VHF and UHF (405 and 625 lines). This necessitated complex switching to handle both standards, and with the valve technology available in those early days of colour, sets gained a history of poor reliability.

Fig. 7.10   Television channels

Band	Channel	Frequency MHz
VHF Band 1	1 to 5	45 to 66·75
VHF Band 3	6 to 14	179·75 to 219·75
UHF Bands 4/5	21 to 69	471·25 to 855·25

When BBC1 and ITV moved onto UHF in 1969, it was no longer necessary to manufacture dual standard receivers, and from then on only single standard sets were manufactured, operating only on 625 lines (usually with only a UHF tuner). In Europe, colour is also transmitted in 625 lines on VHF Bands 1 and 3, so European sets all have both VHF and UHF tuners, and some find their way to this country. Although there are still a few dual standard sets somehow existing, the theory is that there should be no more 405-only sets still operating, these being over 15 years old and (presumably) totally finished. Unfortunately, the UK Government does not yet have the courage to switch off the 405 line transmitters which are still being operated and maintained at great expense by the broadcasting organisations, in case a few thousand people lose television. Of course, it would be far cheaper to give each consumer a new television than continue the present unnecessary duplication. So, if offered a second-hand television in the UK, ensure that the UHF section is operating, i.e. that you can receive BBC2, otherwise you might find yourself without television within a couple of years.

8   Teletext and Viewdata

Angus Robertson

Did you know that in addition to transmitting normal television programmes, both BBC and ITV originate additional news, sport and
information services under the generic name teletext? Since 1973,
these services have been received by all modern three channel television receivers, and only require a special decoder to enable the
hundreds of pages transmitted to be viewed. Over the past two years,
the British Post Office has been developing a similar, but more versatile, service called viewdata. Just to confuse matters, each of the three
originating organisations has its own trade name – Ceefax, Oracle
and Prestel (Fig. 8.1). The two different means of transmission for

**Fig. 8.1   Originating services and trade
names of services**

Generic name	Service name
Television	BBC ITV
Teletext	Ceefax (BBC) Oracle (ITV)
Viewdata	Prestel (Post Office) Others

teletext and viewdata are shown in Fig. 8.2. Basically teletext is added
to normal television pictures and can be received, free of charge, by
any TV viewer (with a licence) equipped with a specially modified
television that can either be purchased or rented. On the other hand,
viewdata is received by calling a special telephone number using the
normal household telephone line. Again, a modified television set is
required, connected to the telephone line by a normal extension plug

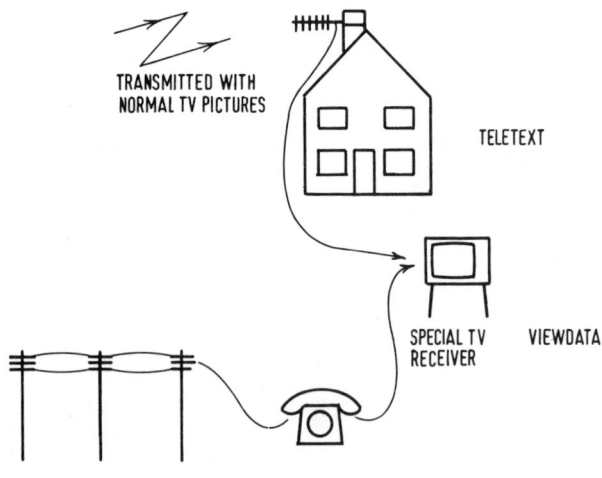

TRANSMITTED WITH
NORMAL TV PICTURES

TELETEXT

SPECIAL TV VIEWDATA
RECEIVER

TRANSMITTED ALONG TELEPHONE LINE

Fig. 8.2 Means of transmission for teletext and viewdata

and socket, to receive viewdata pages for which, unlike teletext, a charge is normally made. Fig. 8.3 shows the major differences between teletext and viewdata which should become available during early 1979 as a public service.

Teletext

Separate research teams at the BBC and IBA developed different versions of teletext named Ceefax (literally 'See Facts') and Oracle (Optional Reception of Announcements by Coded Line Electronics) during 1973. Since there was little point in having the two broadcast-

Fig. 8.3 Comparison between teletext and viewdata

Feature	Teletext	Viewdata
Transmission system	TV pictures	Along telephone lines
Number of pages	Few hundred	Potentially millions
Max. time to receive page	Up to 30 seconds	About 2 seconds
Cost per page	All free	Few pence (some free)
Running cost	Nil	Local phone call charge while connected
Availability	5 years old	Early 1979
Receivers	Widely available	Very few
Future potential	Small	Very considerable
Interactive	No	Yes; message services, games

Both services require purchase or rental of a special television receiver with decoder.

Fig. 8.4 Ceefax teletext page displayed on Finlandia rental Granada set

ing organisations transmit services requiring different receivers, the best points of each were combined to give a 'unified teletext specification' subsequently adopted by the BBC, ITV and most leading European broadcasting organisations with the notable exception of the French who have (just has happened with colour television 10 years ago) developed a completely non-compatible system called Antiope. Teletext has been transmitted daily since 1973, with the BBC producing fully updated magazines of information on both BBC1 and BBC2 for over three years (Fig. 8.4) while ITV introduced its fully updated service two years later.

Teletext is rather cunningly transmitted on four spare lines above the normal television picture where they remain unobserved on a normally adjusted television set. If the picture height is reduced, they can be observed as lines of twinkling dots – eight dots comprise each character eventually displayed on the screen after decoding. Decoders in early receivers were extremely complex with large numbers of integrated circuits to unscramble the digitally encoded teletext signals, and store and display the resulting page of information. Recent sets make use of specially developed integrated circuits enabling much

Fig. 8.5 Ferguson teletext receiver

cheaper and more compact decoders to be manufactured. At the time of writing, teletext television receivers cost between £500 and £700 to purchase (Fig. 8.5), or can be rented from the major high street rental chains for about £16 per month. ITT also market a printer producing a hard copy of particular pages (Fig. 8.6).

Each of the three teletext magazines available (BBC1, BBC2 and ITV) has over 100 different pages that can be selected separately using a small numeric handheld keypad much the same as a calculator. The pages are transmitted cyclically by a computer, each page taking a quarter of a second, so a typical magazine requires 25 seconds to be fully transmitted. Of course, this is a maximum access time for a particular page, and is typically less. ITV transmit rather more pages than the BBC, but use a rather more complicated transmission cycle that ensures more important pages (such as indexes, news and

Fig. 8.6   ITT teletext/viewdata page printer

weather) are transmitted more often than other less important pages such as entertainment.

So what information is actually transmitted on teletext? The single most important topic common to both BBC and ITV services is up-to-the-minute news (Fig. 8.7). It takes approximately three minutes to write and introduce a new page to a teletext magazine (in some cases even less), so the news pages are potentially current to minutes and available throughout broadcasting hours at less than 30 seconds notice. Television and radio news bulletins are transmitted at no less than hourly intervals while newspapers are distributed in each area only twice daily. Typically, some 40 pages of general and financial news (Fig. 8.8) are available on each service with a further 20 pages of sport. As one example, Ceefax has a page for each race meeting with results being added within about five minutes of each race finishing. Thus a complete results section is built up during the afternoon, some 15 hours before an edition of the newspaper *Sporting Life* appears on the street. Football results are also complete by five o'clock on Saturday afternoon, enabling those that miss the televised classified results services to catch up any time during the evening.

Newsflash is a particularly useful service since, when selected, they

CEEFAX 102   Tue   5 Oct   11:11/30

Mrs Margaret Thatcher has warned the
Conservatives, whose Party Conference
starts today, to get their election
machinery ready.

She said an election could come at
short notice.

Delegates at Brighton will be debating
four major topics today: the economy;
education; Northern Ireland; and
immigration.

CONSERVATIVE PARTY CONFERENCE - pp190/1

Fig. 8.7   Ceefax news page

Fig. 8.8   Ceefax finance page showing exchange rates

124

automatically appear superimposed over a normal television programme 'as they happen', typically within a couple of minutes of an
event happening. Ceefax viewers received the news of Elvis Presley's
death minutes before the radio announcement, and half an hour
before the television newsflash. Weather also constitutes a strong position in each magazine, as does travel news about roads, rail, air and
sea. During the summer, Ceefax transmitted a page outlining British
Airways standby seats availabilities to the USA for the following day
– a perfect example of an updatable information service, and interest-
ingly it became apparent that travelling to Boston and Washington is
difficult most of the week and virtually impossible at weekends, while
Chicago has seats almost every day. Ceefax is updated from 7am until
midnight seven days a week, and so always brings up-to-date news and
information.

Other important sections on Ceefax and Oracle are entertainment
information such as television, cinema and theatre, food guides, places
to go (concerts and exhibitions) and fun pages. Although teletext is a
free service to the viewer (except for the special receiver), it has the
severe limitation that adding more pages of information to the transmission cycle increases the access time such that one soon becomes
bored waiting for pages to appear. There are measures available for
increasing the pages two or three fold to perhaps 1,000, but this would
still be totally inadequate to provide, for instance, a respectable classified advertisement service. With this in mind, and for various other
commercial reasons, the British Post Office developed viewdata.

Viewdata
In Britain there are some 23,022,000 telephones connected to
14,862,000 lines (allowing for PBXs and extensions) and the capital
cost of these, together with cabling to exchanges, is £1,548,100,000.
The average number of daily calls per line is only 3·36, and obviously
if subscribers can be persuaded to make more calls each day, this investment will be better utilised thus effectively keeping phone charges
down. The speaking clock is one such 'traffic generator' receiving
around 400,000,000 calls each year: the Post Office hope that their
Prestel viewdata service will be another.

Information is stored at the viewdata centre in much the same way
as teletext using a small computer and bulk disc storage. However
with teletext each page is accessed and transmitted every 30 seconds
or so, but viewdata pages (Fig. 8.9) are only transmitted when a viewer

makes a specific request which makes it a considerably more efficient system and able to handle many millions of pages rather than teletext's few hundred. More importantly, since a telephone line must be established to the viewdata computer to receive information, the computer can then accept other instructions from the home terminal making it totally interactive – in effect it means a commercial computer terminal in ones living room or office. However unlike a commercial computer terminal, the protocol required to communicate with viewdata is very simple and can be understood even by children after only a few minutes' instruction.

Perhaps I should make the rather important point that unlike the free teletext service, charges are made for accessing viewdata pages and these can accumulate rapidly while one browses through the system. One other important difference is that while teletext information is originated by the BBC and ITV, the Post Office is only offering its Prestel viewdata service as a 'medium' in much the same way as the telephone system. In other words, private companies and organisations can rent viewdata pages in much the same way as ordering a

Fig. 8.9   ITT viewdata set displayed 'Exchange & Mart' page from Prestel

telephone, and then transmit whatever information they so wish – the Post Office will originate no pages itself, other than indexing, although certain departments may rent pages themselves for various services such as directories, dialling codes, postal charges or whatever. Obviously the normal laws of libel and decency apply, but pages will not be vetted as such by the Post Office. However, subject matter should be within a specific sphere in order that indexing can be provided leading toward that information.

This might appear rather complex, but in practice is very simple and the system is designed to allow rapid access of information without recourse to directories. Basically, each page provides prompts leading toward other pages in the form of a maximum of 10 choices for further information which are then simply accessed by pressing a single digit on the receiver keypad. This new page should in turn provide further prompts until the required information is reached – the final page has a prompt leading back to the route's beginning so one can start again.

In the initial stages at least, Prestel will be primarily providing hard information such as classified advertisements, houses and cars for sale, stockmarket results, entertainment guides, holiday and travel timetables, hobbies and pastime, cars, education, agriculture and farming, and employment. In addition to information openly available to all users of Prestel, closed user groups can also be created. This is of particular interest to companies and organisations that need to supply rapidly changing information to a large number of outlets widely scattered around the country (or world) such as travel agents, banks, building societies, chains of shops and wholesalers (Fig. 8.10), service departments and so on. In this case, only preselected users will be permitted access to certain pages – these users might for instance have to pay an additional service charge or simply belong to a particular group of companies.

When Prestel is eventually fully operational, the interactive capability will allow information to be passed back from user to the original information provider. For instance, if one has been examining airline reservations, a booking could be completed through the Prestel system and potentially could also be paid for using a credit card (whose number Prestel could conveniently also store). Similarly one could complete a sales transaction and potentially the system could be on-line to banking computers enabling one to examine up-to-the-minute statements and make transactions directly without paper of

Fig. 8.10 Typical business terminal from GEC displaying British Library page

any form. Prestel can also be used for tasks such as calculating mortgages which require considerable effort on most pocket calculators. One of the first Prestel pages demonstrated in 1976 were interactive games although these are not being initially provided on the public service. The Maze Games provided mazes of seven different complexities and one has to move a marker around using four digit keys – virtually impossible with the most complex!

Maybe I should say here that eventually there might be many viewdata services operating both in Britain and around the world. The PO Prestel service is just the first and the PO certainly does not have a monopoly over users of its telephone lines. Mullard Research Laboratories have been privately demonstrating their own viewdata system that uses a radically different indexing system, and eventually this might be made available to the public. Since one of the principle economic necessities about viewdata is that it should be available countrywide for a local telephone charge thus requiring a wide network of computers or lines, the PO will have an initial advantage since many different organisations will be using the same computers and lines.

It might be interesting to look at the charges being made for the BPO Prestel viewdata service. Users will either buy or rent a viewdata receiver from their local TV shop and this will be plugged into a normal PO jack socket as used for extension telephones. Actual receiver cost is difficult to estimate but will probably initially be about £700 or £18 per month, eventually reducing. To contact Prestel, selecting a button automatically dials a special telephone number. The computer answers automatically with a request for the receiver's user number which is stored electronically. When this has been received by the computer, it offers the first indexing page and charging commences.

Apart from the local phone call which works out at 22½p per hour in the evening and at weekends, there is a 3p per minute charge while connected to the computer. Some pages will be free such as indexing and advertisements, while others will carry a charge which is prominently displayed in the top right corner of the screen. This varies between ½p and about 10p although higher charges are possible for valuable information. Most information works out at around ½p to 2p, with only topical news items and such like costing extra. This charge per page is paid directly to the companies that supplied the information less a 5% handling charge for the PO. In addition, information providers as they are termed, pay additional charges for having their information stored on Prestel in the first place. The 'A' rate applies to regularly updated information and comprises a £4,000 a year service charge with a £4 per frame charge on a one year contract reducing to £2,400 a year and £2·40 per frame on a five year contract. The 'B' rate applies to more archival information that is only occasionally updated for which an annual charge of £1,000 is levied with a £1 per frame storage charge – however ½p is deducted from each frame access to cover the increased cost of bulk storage. This second rate is appropriate for encyclopedias which might require about 200,000 frames storage and for which a greater waiting time than the normal one to two seconds might be acceptable so that frames could be retrieved from tape or some such system.

Message services will also become available so that users can send written messages to each other. Prestel is even going to be linked to the international telex network so that users will be able to send messages to any telex users worldwide and vice versa of course (useful for a small business that cannot justify the annual rental for a telex).

Teletext/viewdata pages

Before examining how teletext and viewdata services are produced and received, it is worth examining in brief the actual page format and transmission techniques used by the two services. Fortunately, the page format is identical and this allows considerable economies to be made in decoders which may share the most expensive parts (the memories) with separate processing systems. Fig. 8.11 shows a typical teletext test page from Ceefax which demonstrates all the various display formats currently in use. It can be seen that the page contains 24 lines, each with 40 characters, the top line being used to display the service name, date, channel, time and most importantly a page number. It can also be seen that a full alphabet in both upper and lower case characters (unlike computer terminals which invariably only use upper case capital letters) with numerals and a number of common symbols.

This black and white illustration is unable to show that seven different colours are provided: white, red, green, blue, magenta, cyan and yellow. Although most of the colours are used for graphics, only white, cyan and yellow are commonly used for text since experience has shown the deeper colours are difficult to read on a television screen

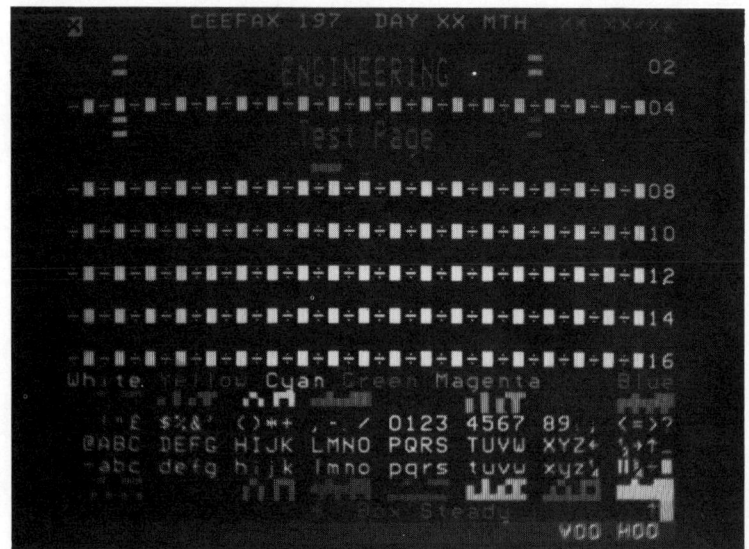

Fig. 8.11   Typical teletext test page showing all display formats

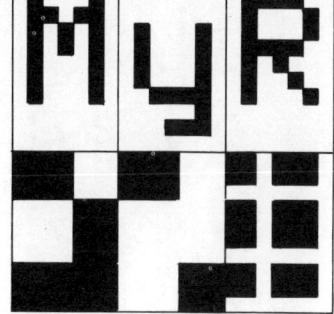

Fig. 8.12 Teletext/viewdata characters are graphics built up from dots on a 10 × 7 matrix

in any quantity. Various graphic characters can also be seen displayed in Fig. 8.11, and Fig. 8.12 shows how these and other characters are built up in a character display 'box' containing 10 by 7 dots. It can be seen from Fig. 8.12 that one square must be left between and below each character to allow for lower case descenders. There are two types of graphic characters, the first two in Fig. 8.12 being contiguous (solid) while the other is non-contiguous (separated) – continuous graphics are best for producing borders, lines and the large display characters, while non-contiguous graphics are more suited to half-tone effects such as graphic representations of photographs.

Although the characters shown in Fig. 8.12 might appear slightly stylised with sharp corners, in practice some system of character rounding is used before these are displayed on the television screen which makes the characters appear more normal to the eye. Other display formats that may be displayed include flashing characters, hidden characters which appear only when a 'reveal' key is selected on the decoder, characters may be displayed double the normal height by using two rows instead of one (a very useful facility for headings and subtitles), boxing allowing subtitles or newsflashes to be inserted within a black box within a normal television picture, coloured background which replaces the normal black background behind characters or graphics with a specified colour (with subsequent aesthetic increase in picture presentation) and graphics hold. This requires the brief explanation below.

The total character set for teletext and viewdata contains 96 different characters and symbols (Fig. 8.13) and is normally displayed in white on a black background on the television screen. However, to cover all the permutations of different colours would mean transmitting 672 different codes for characters alone, with another 672 codes

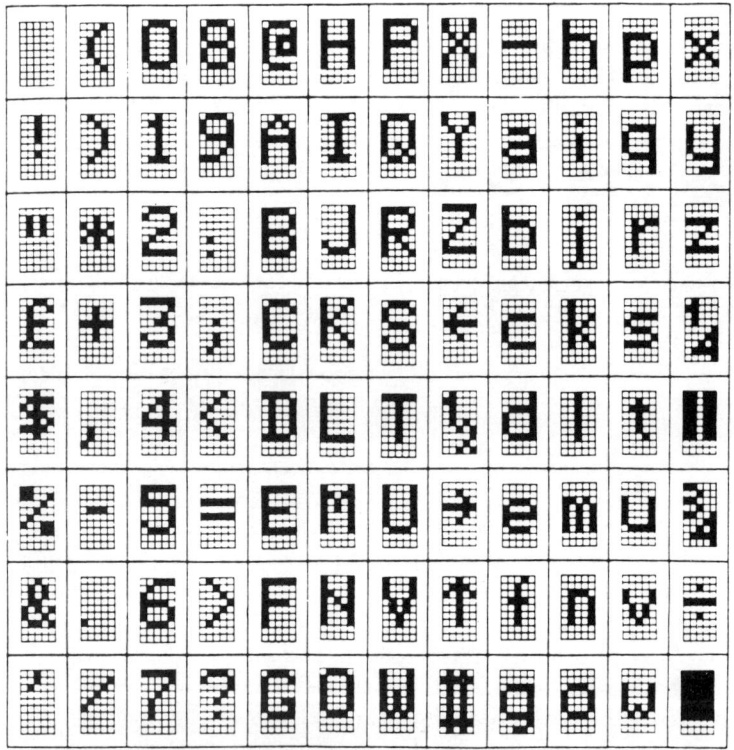

Fig. 8.13 Characters displayed on teletext/viewdata receivers showing how they are constructed from the dot matrix

for graphics. Instead, there are 32 control codes and the 96 character which thus total 128 and these control characters, which are normally displayed as a space on the screen, change the display format in some way. For instance, one could change all the following characters into red (or any other colour), or produce graphics characters of a specified colour. Thus although a vast number of character codes are saved, pages contain many control characters – although displaying them as spaces normally has little effect on the displayed picture since they tend to be used at the beginning of lines or between sentences, where solid areas of colour join each other, a space normally has to be left and this can look rather unusual as in the weather map (Fig. 8.14). The graphic hold facility allows the previous graphics character to be repeated over the space normally left by the control character with resultant improvement in display presentation (Fig. 8.15). There is a slight complication with the double height, background colour, graphics hold and non-contiguous graphics modes in that these facilities (often termed the 'new facilities') were not provided on earlier teletext receivers, many of which are still sold or hired. These older receivers will display double height characters as single height, and ignore the other control characters. The easiest way to check when purchasing a teletext receiver is to select the 'test page' (find the page number in the index) which should look something like Fig. 8.11 with double height characters, solid changing colour graphics and so on; if these displays do not appear on the test page, the receiver will be unable to display them.

Teletext/viewdata receivers

Since both the teletext and viewdata services were designed to be available to consumers as economically as possible, greatest possible use is made of the domestic colour television receiver. Fig. 8.16 shows a schematic of a typical television receiver providing both teletext and viewdata display facilities. The blocks of circuitry found in all colour television receivers are shown along the top of Fig. 8.16, those re-

Fig. 8.14   Weather map displayed using old format

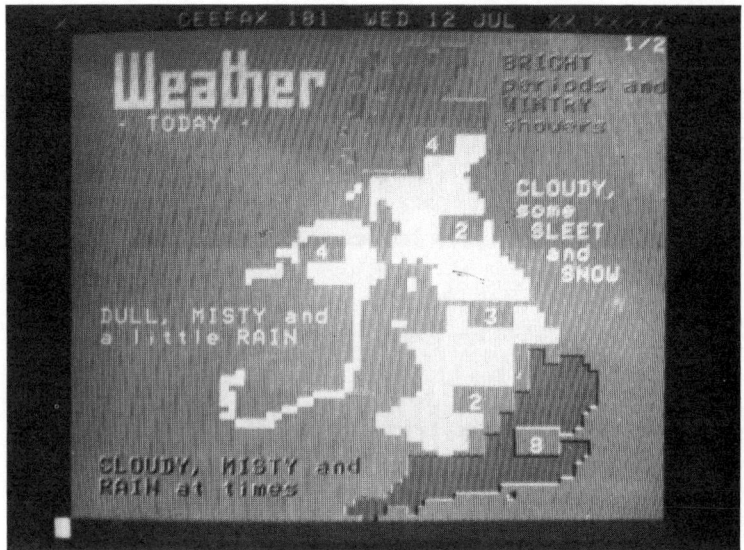

Fig. 8.15　Weather map displayed using new facilities

quired for teletext below, and viewdata blocks in the lower left hand corner. It will be noted that a considerable part of the circuitry is common to both teletext and viewdata with resultant cost savings. Although consumer receivers containing viewdata will definitely also feature teletext (but not the other way round), there will be a market for business viewdata terminals for which neither normal television programmes nor teletext will be required, and these will omit the TV tuner, IF and detector, colour decoder and teletext processing blocks from Fig. 8.16.

In the case of teletext, the pages are transmitted one character row at a time, one complete row of characters being transmitted on each television line. Thus there are four rows transmitted with each frame and 100 rows each second, or just over four complete pages every second. Unless these rows can be identified as being associated with a particular page, it would be impossible to display them correctly. So each page has a 'header' containing a unique three digit page number (and the date, time, etc.). Since teletext pages are transmitted cyclically, a page number entered in the teletext decoder is continuously compared with the page number in the header – when both are identical, the teletext processing block assumes that this is the required page, and loads the following rows (identified as two to 24) into the page store.

During transmission, any interference to the television signals, will result in the digital teletext signal being corrupted and incorrect characters being displayed on the screen. To provide some form of safeguard, all the characters are parity protected so that if a single error is introduced during transmission, then the character is not displayed at all. However, if two errors are introduced into a single character bit, then an incorrect character will be displayed. Since incorrect page and row numbers would cause large numbers of incorrect characters to be displayed, so these are more heavily protected by a technique termed Hamming Codes which allows a single error to be actually corrected automatically, while two errors cause that page or line not to be displayed. Unfortunately, you never get something for nothing, and Hamming codes take about twice as much transmission time as single parity, so are not used on the normal characters.

Once a teletext page has been loaded into the page store, it is ready for display on the television screen. The output processor accesses the page store which is converted into displayed characters using the character memory. The various control codes are also interpreted by the output processor to create boxes within the television picture, flashing characters and so on, and particularly generating colour in the picture. Referring back to Fig. 8.16, it can be seen that the received television picture passes through a colour decoder which produces

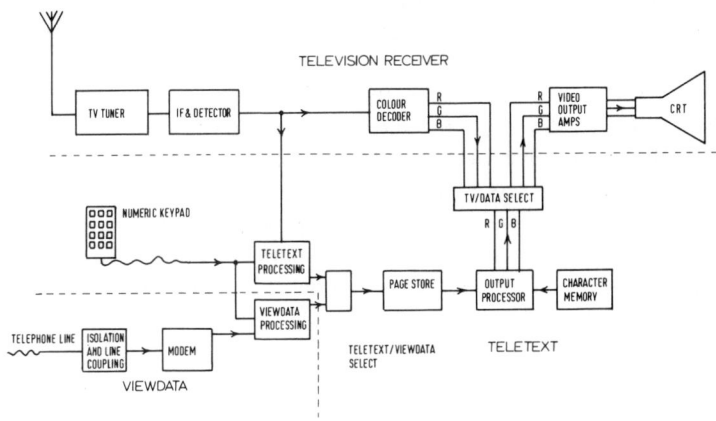

Fig. 8.16 Schematic arrangement of TV receiver showing additional circuit blocks required for teletext and viewdata

separate red, green and blue colour signals to drive the three electron guns in the cathode ray tube. When teletext is added, the displayed characters are added after the colour decoder which means that their quality is better than could possibly be observed from a transmitted television picture which would have an encoded colour signal. The TV/data select block is an electronic switch that feeds either television pictures or data signals to the video output amps and cathode ray tube which displays the resultant pictures.

Having considered the teletext transmission aspects, let us now examine how viewdata differs. It has already been established that viewdata is transmitted along a normal telephone circuit from the viewdata computer centre to the television receiver viewdata decoder. Although viewdata, like teletext, is transmitted as a digital data signal, the telephone line is unable to transmit the signal in this form, so it must first be modulated onto an audio carrier using a modem (modulator/demodulator) and then demodulated in the decoder back into digital information for the viewdata processor (Fig. 8.16). The Post Office Prestel viewdata service uses the Datel 600 standard for the modems, and this provides for the principal information channel to operate at 1,200 baud (bits per second), while providing a much slower return channel at 75 baud.

Although viewdata uses the same character codes as teletext, slightly more bits are transmitted. Each teletext character is composed of seven bits with one bit for parity, making eight in total, but viewdata adds a start and end bit, making a 10 bit 'package' to be transmitted. Thus at 1,200 baud, 120 characters are transmitted each second, producing a potential eight second transmission time for each page. Generally, pages do not include all characters, so the transmission time is shorter, typically five seconds. Although this time is considerably longer than that taken for teletext (quarter second), this is immaterial since viewdata pages are only transmitted on demand. The 75 baud channel is used by the decoder keypad to return information to the computer such as the page numbers required. This return channel can handle $7\frac{1}{2}$ key depressions per second, a speed it would be difficult to achieve in practice.

There are a number of other functions required of the circuitry in the viewdata decoder, the most important of which is isolation of the very high voltages used in the television receiver from the telephone line which is regularly handled by engineers who must not be subjected to such lethal voltages. The line coupling unit also allows the telephone line to be held without the necessity of leaving the phone off the hook, provides an auto dialler to dial the viewdata centre, and gives an automatic identifier that upon enquiry from the computer returns a code allowing the computer to positively identify the terminal calling, enabling proper charging records to be kept – much like the answerback on a telex machine. Once received by the viewdata processing block in Fig. 8.16, the data signal is dealt with identically to that described for teletext.

The display format for the Post Office Prestel viewdata service differs to that of teletext in that the top line of the page has the information provider's name on the left, and the page number in the right hand corner. Unlike teletext, this can consist of up to nine digits because of the considerably greater storage capacity of viewdata. The bottom line cannot be used for page display. Instead it is used to show the number entered by the keypad during page selection and single-line messages from the computer when errors are made. Although a viewdata terminal can gain entry to the database easily, there are certain areas that are restricted to certain users only and which require a pass word to be correctly entered before access is given.

There are a variety of different keypads that may be connected to the television/teletext/viewdata set. With earlier sets, a simple cable connected keypad was used which only provide page selection. With more modern receivers, either ultrasonic or infra-red is used to provide a cordless connection for the keypad, which often also offers television programme selection, autodialling of viewdata and control of sound and picture levels on the receiver. Although for normal information retrieval on viewdata, only a simple 0-9 with * and # signs are required, if the message services are to be used, a full alpha-numeric keyboard may be easily added.

Although the most economical teletext receivers will always be those with built-in decoders, there are still over 20,000,000 ordinary television receivers in Britain, and not all households wish to purchase a totally new receiver with teletext built-in. Thus three companies have developed stand-alone teletext adaptors that can be plugged into the aerial socket of a normal television receiver. Fig. 8.17 shows a schematic layout of such an adaptor, and it can be seen that it contains many of the modules used in the normal teletext receiver, with additionally a colour encoder and RF modulator. Although such adaptors provide teletext on a normal television receiver, the picture quality is much lower than built-in decoders due to the colour encod-

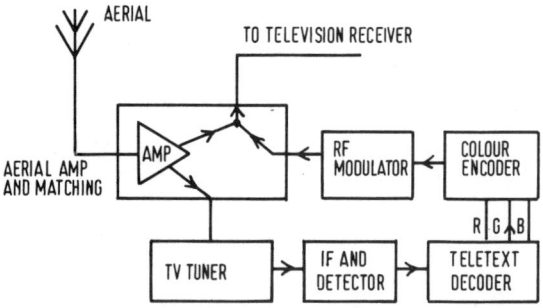

Fig. 8.17 Schematic typical add-on teletext adaptor

ing and modulation process. Labgear manufacture the Colourtext for about £400, while Radofin (Fig. 8.18) and Teleng both market adaptors for less than £200 and these are unusual in that they include all the new facilities.

Originating centres

So far, only the receiving end of teletext and viewdata has been considered, but the originating centres also bear examination. Teletext is broadcast with normal television programmes, so it is obvious that both the BBC and ITV will have origination centres. Although it is eventually intended to operate separate regional services, currently both BBC and ITV operate national services, the ITV service being produced jointly by London Weekend and Independent Television

Fig. 8.18 Radofin teletext adaptor selling for less than £200

138

News. The teletext editing centre used by each organisation is similar to that shown schematically in Fig. 8.19. Basically, there are a number of editing terminals which comprise a keyboard and display monitor, entering information into the main computer, which is then stored on cartridge discs with a capacity for around 1,000 pages of information. These are cyclically accessed by the computer, turned into teletext format signals by the data encoder, and added to the television signal going to the transmitter by the data inserter. In some cases remote keyboards are used, with modems, for instance enabling the BBC sport and financial information to be generated in Broadcasting House in Central London, while the actual Ceefax editing centre is in Television Centre in West London (Fig. 8.20), some five miles away.

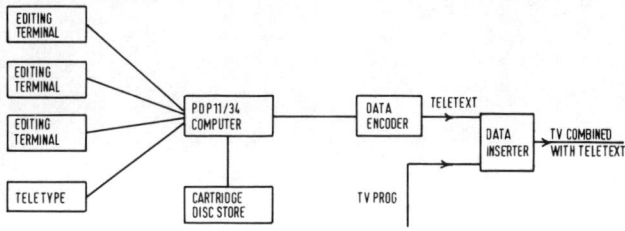

Fig. 8.19   Schematic teletext editing centre at television station

On the other hand, ITV have complete computer centres in LWT (Fig. 8.21), Thames and ITN, all being interconnected for transfer of pages. When regional teletext editing centres are set-up, it is likely that the centres will only insert information on certain pages, leaving the national information intact, and similar arrangements can be made in hotels or conference centres for instance, where the management might care to add information such as room services to the teletext magazines broadcast nationally.

Although both BBC and ITV employ qualified journalists in their computer centres to produce the broadcast pages of information, the PO Prestel viewdata service supplies information supplied by almost 200 independent information providers spread around the UK, and so a different technique is required. Fig. 8.22 shows (schematically) a Prestel viewdata centre which uses a GEC 4080 computer with a 70 Mbyte disc memory capable of holding 70,000 pages. Although this configuration is similar to the pilot Prestel centre, those for the public service will contain a main computer with four 70 Mbyte discs, and a

Fig. 8.20 BBC Ceefax teletext editing centre showing keyboards where pages are entered into the computer

Fig. 8.21 LWT Oracle teletext editing centre

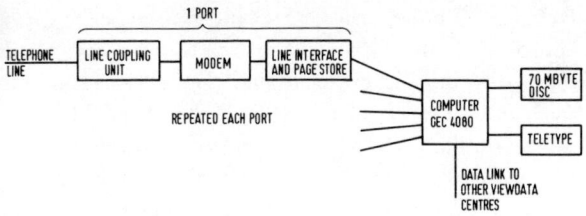

Fig. 8.22 Schematic for viewdata (Prestel) centre in telephone exchange

standby processor with twin 70 Mbyte discs, providing 96 and 48 ports, respectively, and all normally in use. Referring back to Fig. 8.16, it can be seen that each port comprises a line coupling unit which provides automatic answering of calls and interfacing to the modem, and then a line interface and page store which provide connection to the main computer. When a request for a particular page is received, this is passed to the computer which locates and transfers the page from the disc to the line page store at high speed, usually within a hundred milliseconds. The computer is then able to proceed and handle further requests, while the line interface sends the page at 1,200 baud back to the requesting terminal. Providing information recall services of this nature requires little computer processing time (which is inherently expensive) unlike messages and games, and so early Prestel services will be so restricted.

One of the objects of the PO Prestel service is to persuade telephone subscribers to make greater use of their telephones, because the line to the local exchange will otherwise be unoccupied; but this is not so for lines between exchanges which are provided essentially to meet demand. In order to keep the new traffic being generated by viewdata relatively local, computer centres will be set-up around the country as demand for the service grows. Eventually most major towns will have their own viewdata centre, but initially in the first stage, computer centres will be located in Birmingham, Edinburgh, London, and Manchester, with Glasgow, Liverpool and Nottingham all having local call access via multiplexers to their other viewdata centre. Since 26% of PO Prestel customers are located in London, a number of centres will be located in different parts of London to cope with the demand. Local call access will be available for Prestel by the end of 1979 in the aforementioned cities. This does not prevent customers in other towns receiving the service, provided they can accept the higher trunk call charges. The second stage should commence in early 1980 and adds computer centres in Cardiff, Chelmsford, Leeds (and more in London), with multiplexers in Bristol, Hemel Hempstead, Newcastle, Norwich and Reading. (To those in the UK Hemel Hempstead might sound an unlikely location for a Prestel multiplexer, but there are a large number of towns around the area that have local call access to Hemel Hempstead, and so will also be served with Prestel.)

During 1980, the PO plans to have sufficient Prestel centres to support 100,000 users, i.e. about 5,000 ports. Stage 3 commences in mid-1980 and will bring Prestel to 55% of the UK users, gradually increasing to 90% by 1982. However, the whole operation is flexible, and if demand occurs in unexpected cities and towns, Prestel can be brought in within months by use of a multiplexer from another centre, a full computer centre being added when demanded.

Prestel itself will be used to determine whether potential customers are within local call access of a computer centre or multiplexer, by simply keying in their telephone number. Naturally, Prestel allows immediate updating of this information, as new centres open.

Information providers mostly have normal viewdata receivers, but with a special editing keyboard in place of the simple keypad that allows the various control codes to be generated. When an information provider wishes to enter information onto Prestel, or update information already there, the user calls the computer normally, and then selects a particular page that leads into the editing system. After supplying a password, they are permitted to update or enter pages under their own 'entry point' into the system. This prevents fraudulent amendment of other information from other organisations. This has been the only system available for entering information onto Prestel since 1975 and has proved to be limiting in facilities, so a number of companies have developed viewdata editing systems. These are not unlike those used for teletext and which provide local processing power and memory and providing considerable time savings in producing pages, they are stored locally during an editing session, and may be transferred at high speed into the central viewdata computer by a special bulk update facility which is a considerable improvement on the two minutes it takes to send a page at the 75 baud rate used by normal editing terminals.

Another device which saves time is a television camera connected to a graphics generator, which is able to produce teletext/viewdata graphics directly from original artwork.

Into the future

Both BBC and ITV are presently (1979) spending considerable sums of money developing on-line systems for the preparation of subtitles for teletext. Currently, the deaf and hard of hearing have only a brief subtitled news headlines each day, with *News Review* each Sunday but teletext allows subtitles to be transmitted and only viewed by those who need them, so they do not 'annoy' the majority of viewers fortunate enough to have normal hearing. Preparation of subtitles is very time consuming and totally uneconomic for most programmes, so computer techniques are being used to simplify the problems and enable this service to be readily provided. One technique uses a phonetic Palantype short hand machine whose output is analysed by a computer to generate captions. Unfortunately, the English language has many multiple meanings for identical sounding phrases, and so the technique is not perfect – but with a little practice, one can easily make sense of the captions.

Other research being performed allows teletext and viewdata graphics to be generated directly from artwork, using a small television camera rather than the present technique of laboriously plotting graphics onto graph paper before entry on a keyboard, one character at a time.

Telesoftware is a joint research project of Mullard and the Independent Television Companies Association, and essentially allows teletext or viewdata to transmit computer software (programs) for a home computer attached to the teletext/viewdata decoder. This is a considerable saving in cost and time over present distribution techniques involving cassette tape and floppy discs.

Possibly viewdata systems will be developed by companies outside the Post Office where the Prestel system is unable to provide certain facilities. Prestel will gradually expand, and begin offering message and telex facilities, with potential interconnection with other databases. Potentially, it will no longer be necessary for bank statements to be posted out monthly, since one could access ones own account directly through viewdata. Take it a step further, and cheques and credit cards will be unnecessary when cash registers in shops and stores are brought on-line to the same computers, providing instant debits without any paper work. This may be a frightening possibility, but it is something that has already undergone experimentation in the USA, and is not many years off in the UK. Like all advances in technology, careful and conscientious monitoring is essential.

Most of the preceding applications have assumed the teletext or viewdata are effectively on-line, be it via a television transmitter or telephone line. Thus neither system could replace newspapers or magazines which are often read on trains or other such places where it would be impractical to arrange either a television aerial (unless signal strengths are increased by at least two orders, 100 times), or a link into a viewdata computer (whether radio or hard wired). Although satellite circuits using rather higher wave bands than in use today with an exterior carefully positioned dish antenna might solve some of these communication problems, this will still be impractical for many mobile applications.

In the more immediate future, even when the cost of computer processing and storage falls, the cost of communicating with viewdata is going to remain relatively high because of the telephone line element. This will remain until the telephone network is totally digital (in the 1990s) and optical fibre cables (with almost unlimited bandwidths) replace the present copper and aluminium conductors which have severely limited bandwidth; bandwidth costs money. Optical fibres will enable very high data rates to be transmitted to the home terminal, in the region of millions of baud (bits per second) rather than the present 1,200 baud of viewdata. A similar rate may be transmitted by teletext if the entire channel bandwidth is used. For instance, during the night, or if a specified television channel were provided (such as will become available when the present incredibly wasteful duplication of television services on both VHF in 405 lines and UHF 625 line colour is discontinued in the UK. It is now over 10 years since the last 405 line only television set was sold. Surely there must be no more than a few thousand homes remaining which are not able to receive UHF television, so the maintenance effort, power and bandwidth being wasted by these VHF transmitters must surely end. It will then be possible to redeploy this valuable bandwidth for communication services, extra FM radio bandwidth, and special data transmission channels that will provide high speed teletext services, with magazines containing millions of pages rather than the present hundreds.

Thus dedicated data transmission channels will provide vastly increased teletext services, while optical fibres will enable much faster communication and be more economical with viewdata. What about the portable terminal aspect? Although the technology for what I am going to suggest is not yet available, it is all at the research stage and should be available within five years. A portable data terminal will become available which will include a solid state liquid crystal (as in watches) alphanumeric display capable of displaying at least 200 characters, and preferably a full 960 character teletext format page. It will provide storage for at least a thousand teletext/viewdata pages (containing perhaps 100,000 words, about one book or a couple of newspapers or magazines). This might be 'loaded' from the home terminal with, for instance, ones selection of newspapers transmitted overnight on teletext and stored in the portable terminal for reading on the train or wherever.

Perhaps a report might be transmitted by 'electronic mail' – a technique not dissimilar to viewdata where office word processors (electronic typewriters) are interconnected by communication circuits and computers, enabling letters and correspondence to be transmitted electronically between terminals and being read on a viewdata type terminal and then produced as a hard copy by a printer in addition to the computer memory files of course. Electronic mail will also replace telex and possibly certain aspects of facsimile (transmission of images electronically). Again, the gradual introduction in the UK of a digital telephone network during the 1980s and 1990s will considerably simplify the attachment of viewdata and electronic mail equipment since modems will no longer be required, and call set-up which presently takes 10 to 15 seconds, will be reduced to maybe two or three seconds.

Although teletext, viewdata and electronic mail will not only provide a dramatic increase in information availability, they will also inherently account for a loss of jobs in the postal services and printing trades with all the associated social problems. It would be wrong not to consider these aspects alongside the development of these systems.

9 Large Screen Television

Angus Robertson

Ever since the 1940s, 26 inches has been the excepted maximum television tube size – although this perhaps arbitrary size came about due to the mechanical problems of manufacturing larger glass envelopes for the tubes (not to mention the weight of larger envelopes), this size has also become convenient for viewing in the average living room at a distance of around 10 feet. On the other hand most people also regularly visit the cinema where picture size is almost invariably effectively larger, even when viewed from a greater distance. Although the normal television tube, more correctly termed a cathode ray tube, can now be manufactured to a maximum of around 32 inches diagonal, if larger pictures are required it is necessary to revert to a projection technique where a smaller television tube is enlarged using some form of magnifying lens – commonly referred to as 'projection television' or 'large screen television'.

Projection television was initially introduced during the 1930s because the manufacture of cathode ray tubes larger than five or six inches proved impractical, so a simple lens arrangement was used to magnify this picture onto a somewhat larger internal screen mounted within the ornate cabinet. As larger cathode ray tubes were developed, this technique faded out. Meanwhile a Swiss professor had developed the Eidophor projector capable of projecting television pictures up to 20 feet wide, with the hope that families would flock to the local cinema to view television programmes. Unfortunately, this never happened since television receivers became a normal household item, and his scheme failed. However the technology has been developed over the years and there are perhaps 100 Eidophor television projectors in use around the world. Since each colour projector costs over £100,000, they are not exactly common. They are used for projecting large television pictures at rock concerts, elections, in large lecture theatres and cinemas – or, typically when a live boxing match is relayed by satellite in the small hours.

Eidophors also find industrial and commercial applications such as aircraft simulators and NASA uses several in Mission Control. Eidophor projectors use a device termed a light valve to scan and vary the intensity of an intense beam of xenon light, one light valve being used for each of red, green and blue colours. Apart from a rack of electronics, Eidophors contain many complex mechanical parts such as vacuum pumps and blowers to cool the xenon light source. So although they are marvellous when a 35 foot colour picture is required, they are totally uneconomic for consumer use. Obviously a total rethink was required, and basically two different types of television projector have emerged – one group using specially developed technology and costing between £2,000 and £5,000, and the second comprising a normal 13 inch colour television receiver mounted in a special box with a large lens to project its image onto a special screen, and costing from £800 to £1,500. Although cheaper, the second type projects a rather dim picture that can only be seen in very subdued lighting conditions.

The four principle television projection techniques are diagrammatically shown in Fig. 9.1, in each case three systems being required for a colour picture except for (d). Fig. 9.1a shows an early technique which uses a Schmidt optical system (as used in reflecting telescopes) where the image from a high intensity cathode ray tube of between three and five inches diameter is directed at a spherical mirror, and reflected back past the CRT (which causes little light loss due to its size relative to the mirror diameter) and onto a distant screen. A correction plate (lens) is used to compensate for deficiencies in this system. The optical 'speed' or efficiency is dependent upon the diameter of the mirror and is equivalent to a lens of similar diameter – but mirrors are somewhat cheaper to manufacture than lenses, so the external (because the various components are separately mounted) Schmidt system is capable of producing relatively bright pictures. On the other hand, since the various components are mounted separately, it is somewhat mechanically complex and subject requires regular and comprehensive alignment.

To overcome this, the Advent Corporation developed the internal Schmidt tube which encloses the various components within the same glass envelope as the cathode ray tube (Fig. 9.1b). A small hole in the

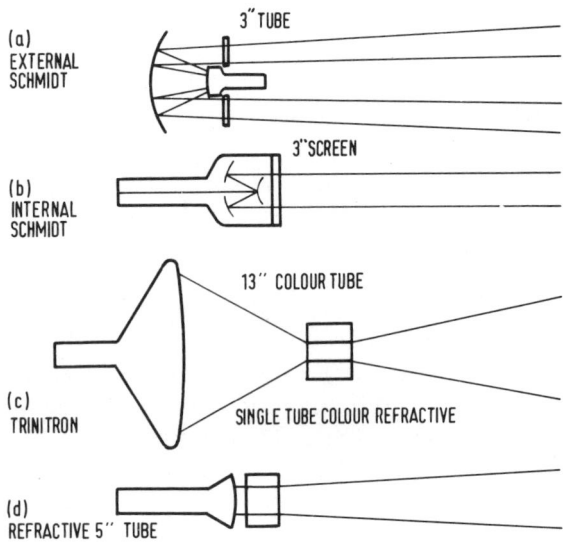

Fig. 9.1 Four projection television techniques

mirror enables the electron beam to scan an internal screen – this is more clearly shown in Fig. 9.2, and the Advent Videobeam 1000A using this technique in Fig. 9.3 with a screen diagonal of seven feet. Three separate tubes project red, green and blue images which are mechanically arranged to converge on the screen exactly 100 inches away. Although this allows simple alignment, screen size is restricted

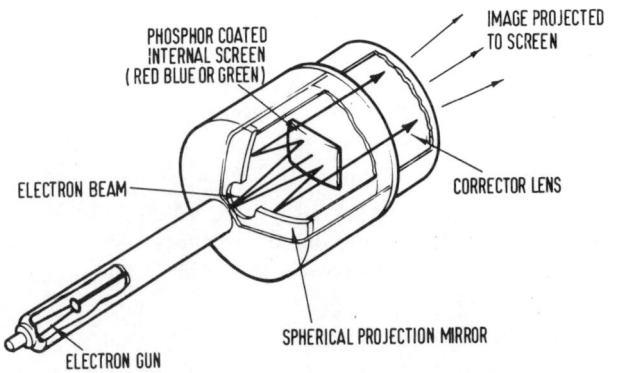

Fig. 9.2 Internal Schmidt optic projection tube

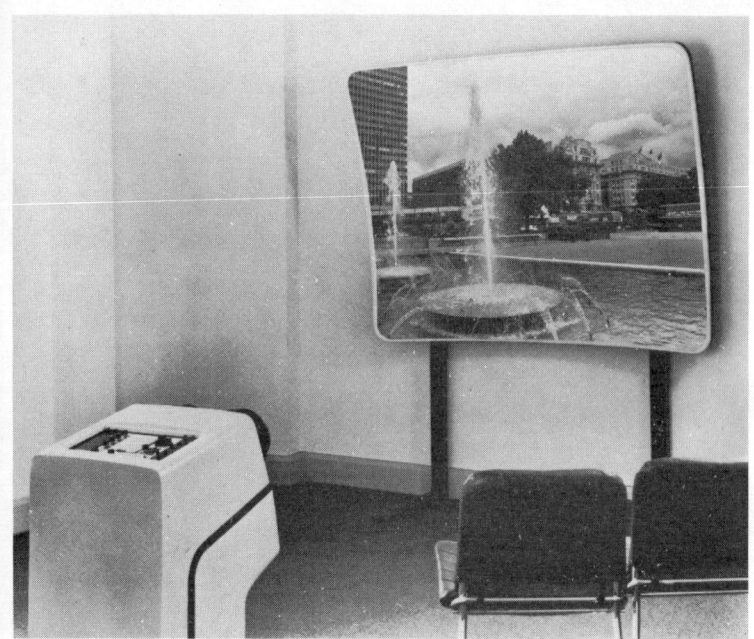

Fig. 9.3 Advent Videobeam 1000A television projector

to this seven feet. Similar projection tubes using internal Schmidt optics are now also being used by National Panasonic and Mitsubishi.

Fig. 9.1c shows the single tube colour refractive projection system. The tube is typically a 13 inch Trinitron television receiver, while the lens is usually acrylic, but sometimes glass is used although this is rather expensive to manufacture in the required diameters of three or more inches. Although acrylic can be moulded into approximately the correct shape, a final optical polish is required which keeps the cost up. Although 10 inch diameter acrylic lenses have been experimentally produced (resulting in superb pictures), commercial manufacture has not yet been achieved. Fig. 9.4 shows the Teletheatre projector comprising a Sony Trinitron receiver with a hood and lens mounted on the front, projecting onto a special four foot diagonal screen. There are perhaps 100 more companies producing projectors of this form in both the USA and Britain. Since the light from a 13 inch screen is now filling a screen of vastly larger area, obviously the brightness is reduced – in an attempt to counteract this, special screen

149

Fig. 9.4 Tele-Theatre Trinitron based refractive projector

materials which artificially increase picture brightness at the expense of angle view, are used. Nevertheless, none of these pictures may be comfortably viewed in bright light, and virtual darkness is really required. However if these conditions can be achieved, results can be acceptable.

Reasons for the inefficiency of the single tube colour refractive projector include the poor light gathering ability of the lenses which are several times smaller in diameter than the tube, and the limited efficiency of the Trinitron cathode ray tubes which has separate stripes of red, green and blue phosphor on the same screen with a grill mask to prevent the wrong electron guns illuminating the incorrect phosphors. On the other hand, the colour picture thus generated is integrally registered – in other words the three colour images are precisely superimposed upon one another and need no further correction once passed through the optical system.

However, it is possible to generate considerably more light from a single colour (with solid phosphor instead of stripes) cathode ray tube, and so this technique is now used by the more modern television projectors (Fig. 9.1d). Such examples are the Advent Videobeam 750 and 760 and each uses three separate optical systems to project red, green and blue images onto the screen. The optical system comprises a high-intensity, five-inch diameter cathode ray tube with a five-inch diameter lens composed from both glass and acrylic elements which projects the tube image onto a distant screen. Although this technique is very simple and thus economical, the lens does not have the power of Schmidt optical systems, and refractive projectors tend to be less powerful with smaller screens. Both the Advent 750 and 760 uses three separate optical systems, the 750 (Fig. 9.5) with the tubes arranged in triangular formation, while the 760 (Fig. 9.6) uses an in-line formation enabling somewhat simplified electronic correction to be used.

Fig. 9.5   Advent Videobeam 750 three-tube refractive projector

Fig. 9.6   Advent Videobeam 760 in-line three-tube refractive projector

Although the usual technique for colour projection is three separate
optical systems (Fig. 9.7) with each system being electronically and
mechanically aligned to project pictures accurately 'converged' upon
one another on the distant screen. Such optical systems tend to drift
with time, and the three images separate causing convergence errors
on the screen. A Florida based company, Big Picture Inc, has developed a new technique which overcomes this convergence problem
and also allows a simple and potentially more powerful lens to be used

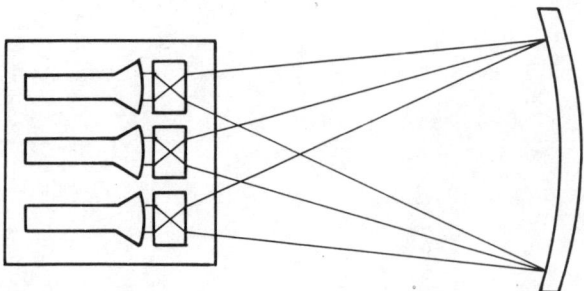

Fig. 9.7   Three-tube colour refractive projector

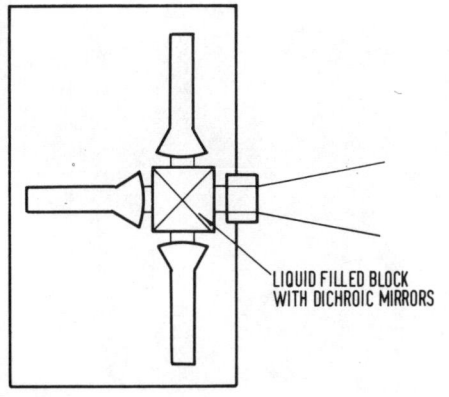

LIQUID FILLED BLOCK
WITH DICHROIC MIRRORS

Fig. 9.8   Aquabeam liquid coupled trichroic colour refractive schematic

(Fig. 9.8). Although the Aquabeam projector (Fig. 9.9) is based on the usual refractive principle, the three cathode tubes have their images combined within a liquid filled block containing two dichroic mirrors. The cathode ray tubes are firmly attached to the liquid filled block which allows brighter pictures to be projected by removing heat from the tube face, and thus allowing higher energy to be used without overheating problems.

Although the earlier projectors used a separate screen, this is generally only suitable for larger than average rooms, and the trend is toward self-contained units with an internal screen. Although the precise optical arrangements vary, Fig. 9.10 shows a typical projector with the projection tubes mounted directly under the screen, and with their images directed at the screen via a front silvered mirror. Rather complex electronic correction is required to compensate for the unusual light path. A number of manufacturers produce television projectors based on this folded lightpath principle, including National Panasonic and Sony (Fig. 9.11).

Screens
Apart from the Eidophor projector, all television projectors provide a light output that is very small when compared to that of a film projector. For instance, a typical 16mm film projector produces about 650 lumens of light output, the Trinitron based refractive projectors offer about 20 lumens, the three tube refractive types from 50 to 80 lumens, external Schmidt projectors about 100 lumens, and finally the efficient Aquabeam from 100 to 300 lumens depending upon the lens

153

Fig. 9.9 Aquabeam television projector

used. It can be seen that in all cases, light output is significantly lower, and so so-called 'high gain' screens are used to make the viewed picture appear brighter. These operate by using a special highly directional aluminised screen surface that provides an effectively brighter picture over a restricted viewing angle. In other words, rather than a little light being spread over a wide viewing angle as occurs with a normal matt screen surface, a lot of light is directed over a restricted viewing angle (Fig. 9.12). Typical viewing angles vary between 30° and 60°, with a severe fall-off outside this angle, and also above and below a normal seated viewing position. Because of the high direc-

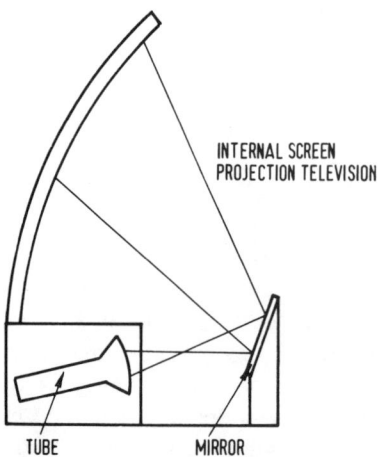

INTERNAL SCREEN
PROJECTION TELEVISION

TUBE   MIRROR

Fig. 9.10 Schematic arrangement of a folded lightpath projector

Fig. 9.11   Sony self-contained projection system

tivity of the screen material, it must be specially curved on a solid moulded base which can be somewhat difficult to transport and store. Screens of this type typically offer a 10 times gain increase over an equivalent flat matt screen. On the other hand there are many applications where these high gain screens are unable to provide acceptable pictures over an audience, and two or more video projectors must be installed.

An alternative to normal screens, are rear projection types. Although these have the advantage of offering a hidden projector since this may now be placed behind the screen, unfortunately it is difficult to design rear projection screens that provide sufficiently high gain for acceptably bright pictures. Nevertheless, developments in this field are expected over the next few years.

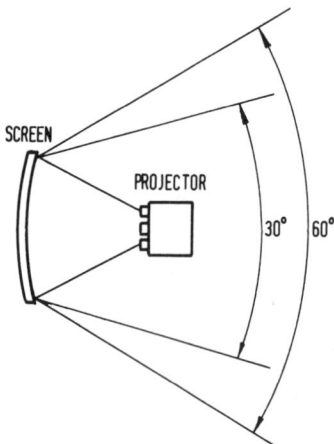

Fig. 9.12 Typical viewing angles using high grain screens

Video cinemas

One particular area where projection television is becoming important is in cinemas, both on the ground and in the air. Although plans for video cinemas have been conceived over many years, most were based on a network of video land lines (cables) linking the various cinemas, and this made the ventures totally uneconomic. With the introduction of video cassettes, a new distribution medium has become available. The financial incentive is substantial. A typical 35mm colour feature film print costs around £500, with a 16mm colour print about half that.

During the past 10 years, there have been many new cinemas opening because of the doubling, tripling and occasional quadruping of screens within the same cinema complexes. Although this means a greater variety of films for the punters, it does increase overheads in terms of the number of distribution prints required for a national release. When the same film is copied onto video cassette, the cost for each 'print' or copy comes down to £20 to £100, and there is also a substantial saving in distribution costs. The average 35mm film 10,000 feet long fills five spools, each about 12 inches diameter. Video cassettes have already been widely used for film distribution over hotel television distribution systems, and now EMI has introduced video projectors into several of its ABC cinema chain, together with video cassette players. The whole system can be totally automated and operated by a programmable timer that also handles house lights and so on. Although such systems will currently be limited to 100-seat and smaller cinemas, developments in television projection should allow larger cinemas to be filled within the next few years.

Airlines have a similar problem with airborne entertainment. Most long distance flights now offer films on sectors of five hours or more, and the same problems on film print cost exist, but multiplied by two or three for a narrow bodied plane, and four or five for wide bodied aircraft since each cabin requires its own film projector. There are three companies involved in this market: Bell & Howell and In Flight Motion Pictures with 16mm projection systems, and Transcom with Super 8mm. Almost without exception, reliability of the projection equipment causes tremendous difficulties often resulting in loss of film in one cabin of the plane. With this in mind, both Bell & Howell and Inflight have originated development programme for suitable video projectors for aircraft. The prototype Inflight projector, developed by Speywood Communication in Britain, is seen in Fig. 9.13 and uses three refractive optical systems with glass lenses, and specially developed cathode ray tubes providing a wide screen format TV picture. The Bell & Howell projector is basically similar (Fig. 9.14) but produces a normal television format picture on a 50 inch diagonal screen. Customers with Bell & Howell projectors include American Airlines, Continental and the private British airline Laker, in all cases the projectors being installed on DC10 aircraft. Bell & Howell/Avicom is now equipping the TWA B-747sp fleet with four projectors each, with a monitor in the First Class. Of particular interest to the airlines is the ability to change programmes while airborne. The film projectors use two foot diameter spools that require careful handling, and certainly not while airborne, so that other programmes such as travelogues can be shown in addition to feature films, and it is also possible to install a cockpit or nose wheel television camera enabling passengers to actually see where the plane is heading!

Into the future

One limiting factor with projection television is the television standard of 625 lines. This was designed specifically for small screen television, and provides a definition limitation for large screen television. This is not important when projection television is used to present television pictures to a large audience, when the screen will be viewed from a considerable distance, but does matter where close viewing is required, such as in the home. With this in mind, the BBC Research

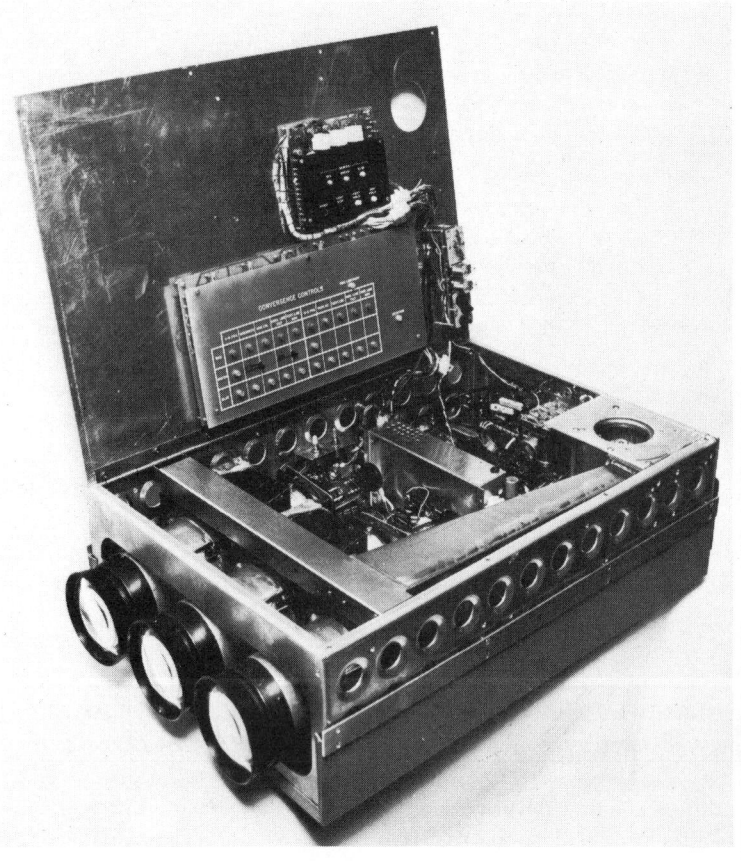

Fig. 9.13 Speywood refractive projector for aircraft

Department has suggested a new television standard providing a widescreen television picture of 8:3 aspect ratio (rather than present 4:3) using 1,501 lines with 2,300 picture elements on each line, and a field repetition frequency of 60Hz on a worldwide basis, even in Europe. This standard produces a television system of 65MHz bandwidth, over 10 times greater than present. It would be totally impractical to transmit such a wide bandwidth on the usual UHF television channels (and even harder to record it), so satellite distribution, probably using digital techniques will be the only way such a service can become reality. Current thinking is that the 42GHz satellite band could offer reliable transmission with a transmitter power of 63kW

Fig. 9.14   Bell & Howell/Avicom refractive projector for aircraft

with digital four phase frequency shift keying modulation. By the time this satellite technology becomes available, large screen television will probably have progressed from the present projection type units to true solid state flat large screen television where the picture is comprised of about 400,000 separate elements in the case of our existing 625 line scanning standard, or about 35,000,000 points for the proposed definition 1,501 line system. Small flat screen television screens have been demonstrated by various research laboratories, and time should bring larger models.

10 Video Programme Production

Barry Hudson

Ask the person in the street today what a video recorder does and you will probably be told, 'it records programmes off your TV'. If you had asked the same person that question five years ago, unless you had hit upon a Latin scholar the answer would probably have been a big 'don't know'. Ignoring the fact that the person in the street in 1978 was wrong (we all know by now you do not record off your television, but rather a television signal from your aerial), what else can a video recorder be used for? The answer, of course, leads us straight into do-it-yourself video. The following advice on the subject stems from experience going back to late 1973 when I had my first introduction to portable black and white video systems. Since then I have made numerous recordings ranging from total disasters to mini-epics on locations (Fig. 10.1) ranging from a Yorkshire coalmine to the top of Brazil's Sugar Loaf Mountain.

Do-it-yourself video falls into several categories. The first consists of a video camera signal fed to a domestic video cassette recorder. This approach is limited by the fact that these cassette video recorders need mains voltage to operate and are unhappy working in anything but the horizontal position, therefore limiting the camera to a few feet away from the recorder (one could use amplifiers, but let's keep it simple). This sort of system can however brings hours of amusement (or embarrassment) by recording various activities at home (Fig. 10.2), and weather and cable permitting, in the garden, the bride leaving for the church, or what have you. The second type consists of a totally portable battery operated system (Fig. 10.3) where pictures are later played back over the home television or transferred onto one of the home video cassettes for long term storage. Pictures of granny on the beach, or some other member of the family competing in some

Fig. 10.1 Typical location recording – here with semi-professional equipment

strictly outdoor activity; pictures of somewhere sufficiently different, the races, a boat trip or holiday, where you'd like to keep a record of the event. Here, the equipment is used rather like its 8mm home movie counterpart but of course with instant results. Both types of equipment are available in either monochrome or colour.

So back to the beginning and let us look at the first type of cameras. All home video recorders, by this I mean Philips, Grundig, Beta and VHS, will accept signals direct from a household television aerial, but some also allow separate video and audio signals to be recorded. To readers contemplating spending money on a video recorder which they intend to use as much for home productions as for recording television broadcast programmes, I strongly recommend they find a competent video dealer to talk to before parting with any money. I say this because nowadays such things as audio dubbing, and tape pause are features not available on all machines, yet they can be significant features in home TV production. For the moment, at least, the advice that follows is aimed at the reader wanting to record, in total, events just as they occur. In mid-1978, monochrome cameras were the answer for someone wanting to take the first steps into tape making, however 1979 saw an enormous reduction in colour camera costs.

One can now expect to come home with a colour camera comparable to its monochrome counterpart, and change from £1,000. However, monochrome still has a role to play so let us consider what is around. Such cameras are available from all the well known manufacturers and some of the not so well known; at the real budget end GBC and Eumig have cameras around £200 and further up the price scale comes National Panasonic, Philips and Sony. The Eumig VC551

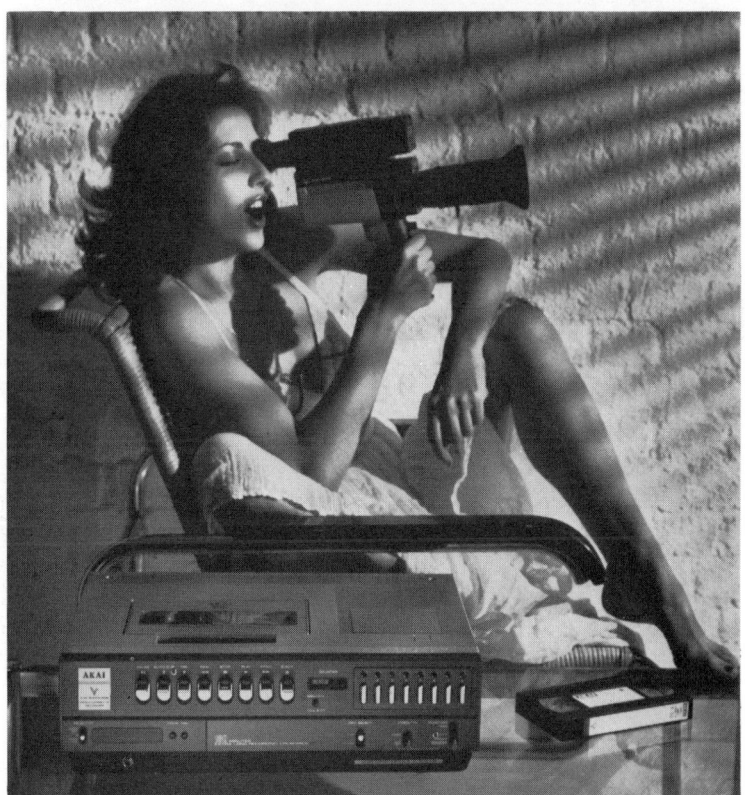

Fig. 10.2 Akai portable monochrome camera with Akai VHS recorder

Fig. 10.3 Akai portable monochrome camera, but with Akai portable VK recorder

has been around for several years and, like the National Panasonic WV460E, GBC Mark 15, and Sony AVC1420CE (Fig. 10.4) uses an optical viewfinder similar to 35mm SLR cameras and the 8mm movie cameras. The Philips V100 however has an electronic viewfinder; this is a small $1\frac{1}{4}$in TV screen fitted into the camera and has the advantage of allowing the camera operator to judge the contrast of the recording. This type of camera is good fun and can readily justify its initial cost, recording parties or golf and tennis swings which need improving. Although dependent upon mains electricity, the majority of the cameras have extension leads so that the camera can be 20 or 30ft from the recorder. However this type of camera is, in the main, equipped with a standard vidicon camera tube. A vidicon is the device which produces the electronic signals which are transferred to and then stored on the video tape, the signals being generated by light falling on the face of the vidicon via the camera lens. The type of vidicon camera tube used in a camera normally dictates the quality of the finished recordings and prices of professional camera tubes can easily exceed £1,000, so the results from a complete camera including lens for under £400 has got to have some inadequacies. The most apparent of these are double and triple image effects, more prevalent under low light levels, known as lag. This can be minimised once the cameraman has mastered the techniques of the slow pan and zoom. Increasing the overall level of light improves the recording but this can be fraught with hidden problems since it is very difficult asking people to relax and act naturally if they are gently cooking under 1,500w or more of light. I can still remember the warm glow of embarrassment that crept over me once at a christening party when someone's grandmother

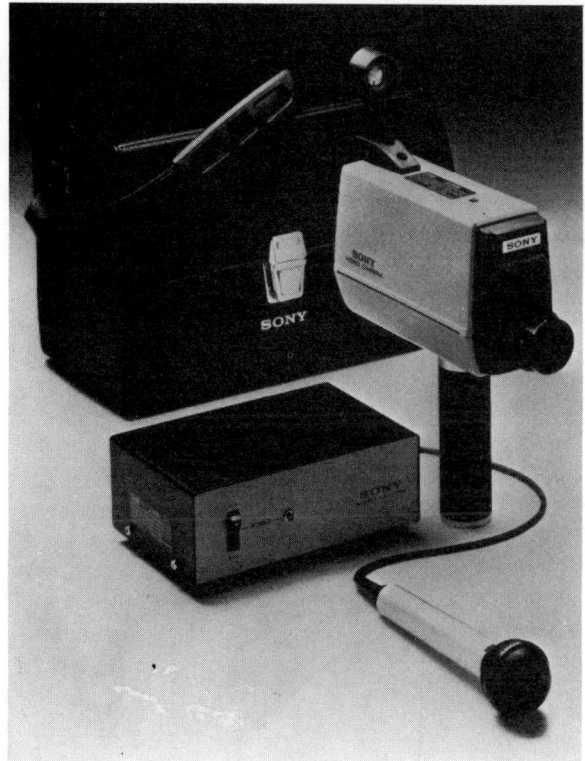

Fig. 10.4   Sony AVC1420CE monochrome camera

made the alternative suggestion, in a very loud voice, about where I might try and put the lights.

Some of the monochrome cameras mentioned have zoom lenses while others have a simple fixed lens, but are at best only 40mm or 50mm at the narrow end of their specification and obviously designed for use indoors. Both the National Panasonic and GBC models have a button which allows them to stop and start certain video recorders by remote control, so this does away with the need to devise a system of hand signals between the cameraman and someone operating the video recorder. Of course, there are several other makes of camera to consider, amongst them JVC, Pye and Sony. In the main, prospective buyers are advised to see this type of camera working.

There are several variations to choose from in the JVC and Sony range, but as this type of camera has become the standard monochrome camera (Fig. 10.5) for use in schools and small industrial set-ups, it is not too difficult to buy very good secondhand ones from the 'small ads'. This type differs from the earlier cameras in being designed to work from a tripod rather than being handheld, and in many ways this is a good thing as it allows the cameraman to set focus, aperture and picture composition – then just leave it running. In fact, it

Fig. 10.5 Sony Portapak portable monochrome camera and video tape recorder which found wide acceptance in education and industry

can give him the opportunity to actually appear on the tape himself; many of the most natural and candid shots have occurred once the video camera is allowed to fade into the background. The owner of a mains-operated video recorder, by spending between £100 and £200 on a secondhand unit or between £250 and £450 on a new camera, can obtain the means of making his own monochrome home productions. I should explain that domestic video is available from two very different sources: there is the normal dealer who stocks video because he has hi-fi accounts with companies that now distribute video recorders; also, there are the pure video dealers who sell nothing but video, and it is this type of dealer who normally has secondhand equipment, a hire department, and authoritative advice.

Unfortunately, video cameras, like their recorder counterparts, suffer from noncompatability and even the standard C-mount lens fixing does not guarantee an instant swapping of lenses. Add to this the fact that some cameras have only video and audio outputs, others only UHF or VHF, or a mixture of all three. Fig. 10.6 gives an indication of how camera and video cassette recorder are linked.

Portable monochrome systems offer a far greater range of goodies to the prospective buyer. At one time or another most established video companies have offered portable video tape recorders for sale. In the main, such systems use a video tape with a running time of only about 30 minutes. The average portable recorder is about the size of an executive-type briefcase, and contain a couple of lead acid rechargeable batteries making them weigh about 14lb. Most of the accompanying cameras are equipped with an electronic viewfinder as opposed to the optical viewfinder found in cine cameras, giving the budding Cecil B de Mille little excuse for ending up with a poorly focused or underexposed recording, because after each shot the operator simply presses the rewind key on the recorder to take the tape back to the start and selects replay. The viewfinder then turns into a playback monitor enabling a full check to be made before moving onto the next shot. The cameras are also equipped with a microphone, so a simple earpiece is all that is necessary if one wants to check the audio recording as well.

Being designed for outside work, the standard camera lens is usually a 6 to 1 zoom, although the Akai range of portables come within a very good 11·5 to 90 (8 to 1) lens. Until recently, all truly portable systems were based on the open reel to reel system (Fig. 10.5) used by the larger audio tape recorders rather than the smaller audio cassette

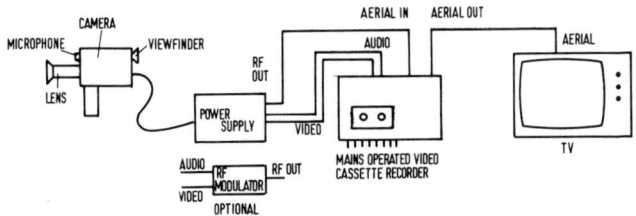

Fig. 10.6 Typical monochrome TV camera connected to normal mains operated video cassette recorder. Some cameras have the power supply built-in, some offer only RF out, only video/audio inputs, then an RF modulator is required in the latter case

format. In common with early tape machines, lacing up the video tape obviously takes longer than simply popping in a new cassette, and the video heads are exposed during lace up. As these are the most expensive and delicate part of any video recorder, great care must be taken during this tape threading operation. In 1977 Akai, which was then distributed in the UK by Rank Audio Visual Limited, produced a portable system using a VK format cassette. The current monochrome model is the VT350 and retails for between £980 and £1,500 (Fig. 10.7). The reason for the variation in price is that the system is available in basic, de-luxe and even super GT versions, so that the prospective buyer can enjoy the almost masochistic experience of trying to choose between what he wants and what he can actually afford.

Sony, National Panasonic, Sanyo and JVC all have portable monochrome video recording systems for sale. It is true that each portable system has its own plus and minus features, but these features only turn plus or minus once related to the known application, so this makes the giving of universally applicable advice difficult. An ideal system for a golfer who wants to improve his game is not necessarily the right system for ballroom dancers who need to analyse their fishtail turn in the quick step. However, in very, very, general terms, someone with £500 to £650 to spend on a fully portable system has a reasonable range of secondhand Akai, JVC and Sony machines to choose from. At the other end of the scale, all the manufacturers are now offering new monochrome portable systems at around the £1,500 mark. My advice is to find a video dealer who is prepared to listen to what you think you want, then ask you lots of questions about application, distances, available light and running times before suggesting a suitable system. Beware the dealer who just reacts to the words portable and immediately produces the only make he has in stock.

Before going on to colour let me explain for the fortunates of this world who are contemplating owning both a mains and portable video recorder, just how the information on the portable can be transferred to your mains video recorder (or for that matter your TV set). Firstly, a video signal requires a device called an RF (radio frequency) modulator which disguises the signal as an incoming TV channel, before your television will receive them and show them on the screen in the normal way. The RF modulator is usually included in portables but is sometimes separate. By tuning in one of the spare channels of your home video recorder and transferring onto its cassette selected shots from your portable, these shots can be kept, thus freeing the tape on your portable for re-use. Fig. 10.8 shows typical connections for a portable and camera. Most of the portables mentioned can be hired from the larger video dealers (once your bona fides have been established) for between £20 to £30 a day which does give the prospective buyer a chance to road test the various models before making a permanent commitment. In many cases, some of the hire charge would, I feel sure, be allowed against the cost of any eventual purchase. The hiring facility also allows the home video owner, by using the transfer method mentioned above, to keep a lasting record of events that took place far from home.

Fig. 10.7 Akai VT350 portable VK format recorder with assemble edit

Colour

If you take a large piece of stiff white card, a box of paints, brush and water, then depending upon your degree of talent, the resulting picture when looked at in daylight could be either colourful and pleasing to the eye or just downright colourful. However, take the same piece of paper down to the cellar at midnight, turn out the lights and that colourful masterpiece will turn into a big grey nothing, put the light on and suddenly all the colours come back. The lesson to be learned is that you cannot have much colour without light. As far as video goes, the cost of the colour camera is inversely proportional to the amount of light it requires, so bearing this in mind let's now move onto colour equipment.

There are many reasons for recording casual goings-on in colour. After all, the majority of our TV viewing is colour, but even simple colour recordings in the home create problems. For example, a dinner party needs careful lighting if the resulting recording is to show anything like true colour. You might spend 60 minutes arranging two or three 800W lamps (Figs. 10.9 and 10.10); you think that everything is ready, the camera is working well and the TV is showing a faithful picture of a perfectly laid table just waiting for the guests to take their places, but when they arrive, all hell breaks loose. For a start, one of the ladies is wearing a diamond necklace which reflects light straight into the camera, then somebody actually has the nerve to use the silver salt cellar, replacing it where it reflects the light and ruins the whole recording.

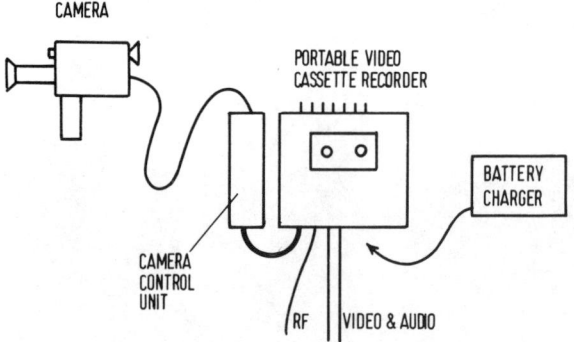

Fig. 10.8 Typical portable system with battery operated video cassette recorder. Some cameras are self-contained, others have some electronics separated, termed a camera control unit as shown here

Fig. 10.9   Location showing two stand-mounted 800W lamps

Fig. 10.10   Another location, this time with two lights suspended from
ceiling rod, and one on a stand

Many of you may by now be thinking it is a waste of time even to consider making your own productions in colour; but actually it is not quite so bad as described above. The main thing to remember is that just because you have spent a lot of money on some pretty complex equipment, you do not automatically become a proficient sound, lighting and cameraman and there is no magic pill to swallow which immediately lifts your first stumbling efforts into the Oscar league. The best advice to my mind – like monochrome – is hire before you buy.

The present state of the colour camera market is very interesting. An article I wrote in mid-1978 advised readers to look out for good secondhand Akai colour cameras around the £1,250 mark. Now, thanks to a range of newer lightweight colour cameras from JVC, National Panasonic and Hitachi Denshi, that advice is old hat – like any other given at a moment in time. Given ideal lighting conditions, all three give very acceptable results. The JVC GC3300 at about £1,450 has a series of knobs on the camera control unit which not only gives a range of colour compensation settings, but like the old Akai VCS150 it allows you to add or subtract red and blue colours from the recording. The Hitachi Denshi GP5 priced at under £1,000, has similar controls on the camera head itself and when used with a suitable recorder, both sound and pictures can be checked out via the viewfinder. So far I have only been able to evaluate an NTSC version of the National Panasonic camera, but, like its competitors, it gives remarkably clear pictures in almost parlour conditions. For those considering the purchase of a semi-professional rather than basically domestic type of colour camera, then the Hitachi Denshi GP7 and FP3030 cameras and the Sony DXC1610P are worth looking at. However, the extra £1,000 in the case of the GP7 and nearly £2,000 for the DXC1610P and the FP3030 need careful justification especially as these cameras are basically designed to work with the ¾in U-Matic format recorders rather than the domestic ½in systems, and you run the risk of using almost too good a camera for the recorder. However, if the end results are going to act as master material and second generation tapes only will be viewed. For example a theatrical manager making demonstration tapes of his stable and then using copies to sell them to the various theatres, will find them well worth considering. At the time of writing this current range of cameras is still too new to have gone onto the secondhand market which only leaves the Akai CCS and VCS 150's, the occasional Sony DXC 1600 and JVC GC4800 to look for, and probably, bearing in mind their original selling price, the present owners will be looking for at least the same as the new range will cost you.

Existing portable systems come in three different, and once again non-compatible, sizes; U-Matic, EIAJ and ½in VK cassette. The ½in cassette system from the Akai stable is the VTS400 and it is unique to Akai. Although the cassette is similar to that of Beta don't try putting it into the Beta machines. The camera uses two vidicons and has the now expected 1¼in viewfinder. The recorder has very neat automatic assemble edit. This is a system whereby, if required, a series of recordings are neatly edited together by the recorder as one goes along, thus doing away with the annoying series of snap, crackle and pops normally observed in playback with other recorders. The 1979 price for the system plus necessary bits and pieces is just over £3,500 but the future may well see a significant reduction in this price of EIAJ standard VTRs; the JVC colour PV/GS4800E is my favourite having an excellent slow motion still frame facility. It is a robust, no-nonsense machine, which if looked after, will give many hours of service. Its weakness (and that only against the very latest models) is that the camera is rather unwieldy and the cameraman is often wishing for a third hand. However, as some of the new generation colour cameras can easily be modified to work with this recorder it is still a model well worth considering. The recorder on its own retails for around £1,250 and secondhand ones must surely exist – and will by the time you read this.

In the U-Matic format, there are two accepted units. The first is the Sony VO3800P/DXC1610P. It succeeds the old Sony VO3800P/DXC1600P (Fig. 10.11) and is the first Sony colour portable that doesn't offer the surgical truss as an optional extra. Frankly, it is what one has grown to expect from Sony, a solid reliable piece of equipment and costs £4,798. Although Sony seems to have suffered somewhat in the domestic off-air market place, it is still the standard outside this area and nowadays anyone wishing to offer an alternative must come up with something a bit special if they want to recoup their research and development costs. Such is the case with the JVC portable U-Matic CR4400E (Fig. 10.12). It will happily accept both Hitachi Denshi's models GP7 and 3030 as well as JVC's own and previously mentioned GS4800E; it differs from the Sony model by being appreciably lighter, a blessing after two or three hours of carrying one around. It also has an automatic assemble edit facility similar to the

Fig. 10.11 Sony DXC1600P colour camera and VO3800 portable U-Matic recorder. DXC1610P is similar, but is self-contained without separate CCU, and rests on shoulder for ease of operation.

Akai system and it is so efficient that one could very nearly make one's own version of a simple childrens TV puppet show such as Magic Roundabout in a first generation tape. It also boasts a very complex audio system which really shines once the recorder is back at base being used to feed either an editing machine for semi-professional work, or when simply transferring the recordings onto a domestic machine. Both the above systems cost £5,571 and £4,882, respectively (in 1978), and are obviously bought with specific requirements in mind. My own company has supplied them for applications varying from a West of England photographer who now makes video tapes of weddings and parties (and makes a feature of being able to play the tapes back over a TV projection screen during the latter part of the

Fig. 10.12　JVC CR4400E portable U-matic and GC4800E colour camera

function!), to an engineering company who use the tapes to help solve overseas development problems with home based management and technical expertise – it must be very useful to see a problem before selecting personnel to be shipped out to rectify it.

The promised range of portable VHS and Beta format machines compatible with mains units are expected to come onto the domestic market complete with camera and a selection of necessary ancillary bits and pieces for under £2,000. By 'bits and pieces' I mean items such as spare batteries, tripod and possibly even lens adaptors, which screw onto the front of the existing lens and convert it to wide angle or even telephoto. The JVC HR4100 VHS format recorder (Fig. 10.13) marries up with its own camera GC4100 and offers close-on two hours of recording from the internal rechargeable batteries. The weight, excluding camera but including batteries and cassette, is under 22lb. It is also rumoured that the Mark 2 version will include features expected on the new HR3600 domestic VHS off-air recording machine including still frame and a double speed play control. Sony's Beta format portable, the SL3000 (Fig. 10.14) is still somewhat shrouded (late 1978), but frankly I would look for a much healthier share of the domestic recorder market by Sony before expecting it to arrive in any quantity.

Fig. 10.13 JVC HR-4100 portable VHS recorder with colour camera

During a recent conversation with one of my own customers, I had the sad task of ruining his day. The disillusionment on his face was total, when I explained that the eagerly awaited VHS and Beta portables are quite limited in what they can do. He was expecting to trade in his present machine for the new model, leave it plugged into the mains recording TV programmes as normal, and then whip off down the golf-course with it to record the latest match via the camera. He could not understand that the portable has had to sacrifice the whole tuner/receiver clock circuitry or 'front end' to make way for the batteries and other bits and pieces to enable one to go 'walk-about' with it, and is therefore strictly a portable machine with no 'off-air' facility. Half-way through my explanation, I suddenly realised that my esteemed and very intelligent customer still thought that his VHS machine was recording from his TV – will they never learn. However, the good news is, or so JVC assure me, that a separate 'front end' complete with timer, designed to bring the HR4100 to full 'off-air' specification is on the way. Price to be confirmed, and arrival scheduled to follow closely on the heels of the portable VHS.

Editing

Recording television pictures on a portable cassette recorder is usually just the first step in making a 'programme' once the initial novelty of seeing ones friends 'on television' has worn off. The next stage is editing and dubbing the recorded pictures – in other words removing all the bits that went wrong, rearranging the order where necessary, removing sound that does not fit, and perhaps adding commentary and music to the final programme. Unfortunately, the equipment necessary to edit and arrange complex dubbing is not cheap, not simple to operate. The basic point is that unlike audio tape, which is 'sliced' simply with a razor blade and sticky tape, video tape must not be cut under any circumstances, otherwise the very expensive video heads are likely to become damaged by the tape edges. Thus video tape editing is electronic. Basically, one copies the selected scenes, with the correct duration and in the right order, onto a special editing video

Fig. 10.14   Sony SL-3000 portable Beta recorder with colour camera

cassette recorder that allows different scenes to be intercut without any noticeable disturbance on the picture. Theoretically, although one only requires a single editing machine plus a playback machine, in practice two editing types operated by an editing controller are used to enable scenes to be precisely timed. Although simple editing video tape recorders were available for the old monochrome formats (and EIAJ colour), these are now all obsolete and virtually all video editing is now performed using the U-Matic format for which a vast amount of development has produced a wide range of equipment – but at a cost. Sony, JVC (Fig. 10.15) and National Panasonic all produce U-Matic editing video cassette recorders which cost from £3,600 to £4,000 each – remember two are required plus another £1,000 for a very simple editing controller to over £10,000 for a rather more complex (and versatile) type. So an editing system does not leave much change from £10,000; a number of companies offer editing U-Matic at from £25 per hour, or £200 a day, and obviously this is rather more economical for most. The same editing recorders also allow audio dubbing to be performed since they include two audio tracks allowing versatile copying from one another, and to other machines.

Editing by its nature means copying, and each copy loses quality. With the U-Matic format, up to three or four copies can be made before the picture severely deteriorates, but the VHS and Beta formats are primarily designed for recording directly off-air which they all do perfectly, but not for copying or editing which, let's face it, very few people would actually want to do. So it is unlikely that any editing equipment will be marketed for either of these consumer formats. On the other hand, it is possible to edit from say Beta or VHS onto U-Matic if the original material warrants it. But the make-shift technique likely to be used by most consumers simply requires use of the pause or still frame control which is fitted to all VHS and Beta format recorders. Although this causes a slight 'bump' on replay, this is not serious on replay unless the video tape is copied again when the disturbance will last a few seconds. Many video cassette recorders also include an audio dub control, which allows the original recorded sound to be replaced by new sound from a microphone. It is not particularly versatile since the original sound will be lost, but it is sometimes useful.

As I mentioned earlier, the simplest editing technique is the automatic assemble edit facility provided in some portable recorders which, each time the pause control is pressed during recording slightly

Fig. 10.15 Typical editing U-matic, the JVC CR-8300 for programme production

rewinds the video tape so that when continuing, a continuous, edited recording is made which may be replayed without any bumps.

In conclusion, I would say that the equipment available to the budding DIY videophile will vary enormously. For those with a limited budget there should be a quantity of second-hand systems made available by those fortunates taking advantage of the new portable VHS and Beta machines; but wherever possible deal with a company which is prepared to give you a lengthy demonstration under 'field' conditions and try hard to get a short-term hiring of the equipment first of all.

11   Television and Electronic Games

Angus Robertson

The past two years has seen the introduction of large numbers of different TV games into the European market. It is estimated that around 1,000,000 games sold in Britain during 1978 valued over £20 million, and is expected to increase in Europe to $300 million by 1985 according to market research specialists Frost and Sullivan. The number of new games being introduced is frankly quite staggering. Technology is advancing at such a rate that completely new generations of games appear about every two years. Unlike calculators, which can be freely manufactured, there are a number of patents covering television games originally taken out in 1966 by the Sanders Corporation, a company developing television displays. It is significant that electronic technology did not allow a consumer television game until 1972 when Magnavox in the USA (now owned by Philips) entered into an exclusive agreement with Sanders and has now licenced about 30 companies worldwide to produce games. The early Magnavox game, Odyssey (Fig. 11.1) was very basic and only featured a ball and two paddles on the screen – there were no sound effects or score and screen boundaries comprised plastic overlays attached to the TV screen.

At about the same time, the first Arcade television games appeared but in this case actual cost was relatively unimportant (several hundred pounds) so they contained vast quantities of integrated circuits enabling boundaries, score and special features such as different ball angles and speeds. Finally, in 1975, General Instruments in Scotland developed the first single integrated circuit television game which included all the features of the earlier expensive arcade games but at a price of around £40. General Instrument has now manufactured over eight million of these integrated circuits which are used by

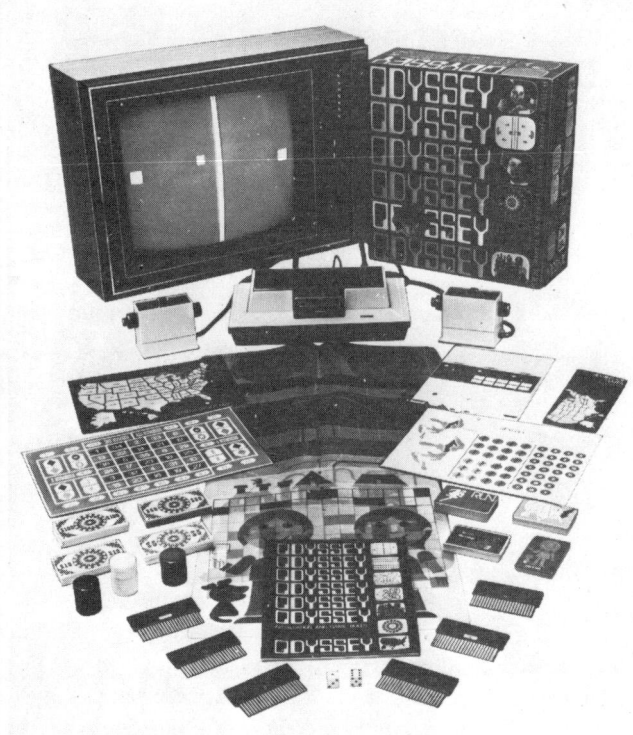

Fig. 11.1 Early Odyssey TV game with many screen overlays, play money, etc.

over 50 manufacturers. Until recently, the only other company that had successfully developed a dedicated integrated circuit television game was Sportel in Luton. This had the distinction of being the first colour game on the market and due to its specially designed integrated circuit, had different games to the other 50 or so on the market – a considerable bonus.

However, over the past year (1979), a number of new games have been introduced both by General Instruments and other manufacturers. These break down into two categories – dedicated types that play a limited number of games such as racing or battles, and programmable units which can play a wide variety of games merely by plugging in different cartridges. Fig. 11.2 gives a brief history of TV games.

Fig. 11.2   Brief history of TV games

Year	Development
1966	Original patents covering TV games granted.
1972	First discrete component bat and paddle game (Odyssey).
1973	Multiple integrated circuit arcade games, with score.
1974	First multiple IC consumer games, no score.
1975	General Instrument introduce first single IC TV game with score.
1976	Numerous Far East manufacturers flood market with games using GI IC. Sportel develops its own single IC game.
1977	Fairchild introduces first microprocessor based TV game using plug-in cartridges for different games. First electronic non-TV game (Chess). Tank and battle games introduced.
1978	Home computers capable of playing complex interactive games such as *Star Trek*. Microprocessor based electronic games such as Simon and Mattel handheld games.
1979	Lower cost programmable games using audio cassettes for memory, lower cost home computers, wide variety of electronic games.

Connection details

Although most games are mains powered, there are some around that operate from batteries. However, these tend to become exhausted rather quickly, especially if left on overnight! In all cases, the game's output is displayed on a normal television receiver by removing the normal aerial lead and inserting instead the game's lead. Inside the game is an RF modulator (a micro television transmitter) that is normally set to between channel 35 and 45 in the UHF (colour) band. The quickest method of tuning is to first ensure the game is operating by turning on, setting auto serve and selecting a game until the unit starts making hit and boundary noises. Then select a spare tuning preselector on the television and rotate the associated tuning control until a strong, clear picture is displayed. It is possible that degraded pictures might be displayed at other tuner positions, but there should be only one 'perfect' setting.

Since aerial sockets on most televisions are not constructed sufficiently robustly for continuous plugging and unplugging, some form of switch/combiner should also be purchased. This will have sockets for the TV aerial lead and TV game lead, and either a socket for a new lead to the television aerial socket or an attached lead. Some combine both inputs continuously, while others have a switch to select the one required.

Dedicated ball and paddle games

Before examining the complex programmable games, perhaps it would be best to look at the present market leader, the non-programmable bat and paddle television games. Although no precise figures are available, it is estimated that about 95% of these games come from the Far East, with 80% from Hong Kong, and the remainder from South Korea, the Philippines, Taiwan, Singapore and other countries. Even games manufactured in Britain by companies such as Teleng, contain many components and modules imported from the Far East. Indeed, Milton Bradley games carry a sticker stating 'Made in Republic of Ireland, integrated circuit made in Singapore'. Even that is not perfectly true, since although the integrated circuit (the electronic heart of all games) is packaged into a plastic 'box', the minute silicon chip containing the actual circuitry is produced in the West (often Scotland or Silicon Valley near San Francisco) and despatched to the Far East for the labour intensive plastic encapsulating procedure.

Most ball and paddle games are based on the General Instrument AY-3-8500 integrated circuit which plays tennis, soccer, squash, practice (squash with one player) and two rifle target games. These latter games use an external rifle with a light sensitive cell and lens system mounted in the barrel which is directed at a moving 'blob' on the screen. There are a variety of options that manufacturers can exercise such as two or four players, random ball speeds, different sized bats, manual or automatic serve, sound output, score on screen and random bat angles. Display is normally white symbols on black background, but can also be grey background with black and white bats for different players (useful in squash), or alternatively can be a fully coloured picture.

The earlier games only had vertical bat movement, so the next development gave both horizontal and vertical movement, sometimes using two rotary controls, otherwise more sensibly with a joystick. Interton manufacture a game that includes grand prix with two racing cars and rallye with one. Other games that have been introduced include a two-player tank battle game, where each player has a steerable tank and a fire control button, and motorcycle which features a single rider moving along three horizontal tracks under the control of a throttle – four different games are provided including jumping buses and a timed drag race.

Although there are a vast number of basic television games available in Britain, the basic market leaders include Binatone, Grandstand (Adam Imports), Videomaster, Radofin, Optim (Optimisation), Ajax (Acme Electrics Co), and Ingersoll (Fig. 11.3). Typical of such games are a vast range from Optimisation Ltd under a number of different brands: Universal, and variously offering monochrome or colour, all with soccer, squash, tennis and solo and some also with shooting, and tank; Spectrum 6 colour games offering the 'basic four games' with three different coloured playing surfaces and randomly moving outfield players; Spectrum 10 with the basic four, shooting, ice hockey, basketball, rugby and offering joysticks with attack buttons; Susie, a basic four monochrome mini-game ideal for children; Combat, another monochrome game but offering tank battles; Popular, a large range of colour and monochrome games, most offering the basic four with shooting, and then variously additionally car racing, tank, stunt cycle, ice hockey, golden shot, basket ball and rugby. Most Optim games operate from six $1\frac{1}{2}$V batteries or a mains adaptor. Among the games offered by Radofin are a hand-held monochrome type with football, tennis, squash and squash practice, or colour game providing football, hockey, tennis, gridball, basketball, target, squash, basketball practice, target practice, squash practice, the game being offered with joysticks for both vertical and horizontal movement.

Videomaster, now more correctly Waddingtons Videomaster Ltd, has been involved with television games since the start in 1975. The current range of bat and paddle games include All Star providing tennis, football, squash and solo in black and white, and accepts an optional Shooting Star rifle for two shooting games. Videomaster Colourscore II (Fig. 11.4) is identical in appearance to All Star but is obviously in colour. In fact, styling and appearance of Videomaster games are to an extremely high standard with cunningly designed packaging that allows the games to be operated with hand controllers stowed in the game's main body, with cables carefully hidden. This facility is lacking on so many games resulting in horrific tangles of cables. Both the previous Videomaster games use rotary controls for shifting players, while the Sportsworld game offers hand-held joysticks and colour display of 10 games including tennis, squash, hockey, solo 1 and 2, football, basketball, gridball and two target games. With the various non-programmable games, target shop prices for basic four game units is £10, colour basic say £15, and six or more games with battles and racing at £20-£25.

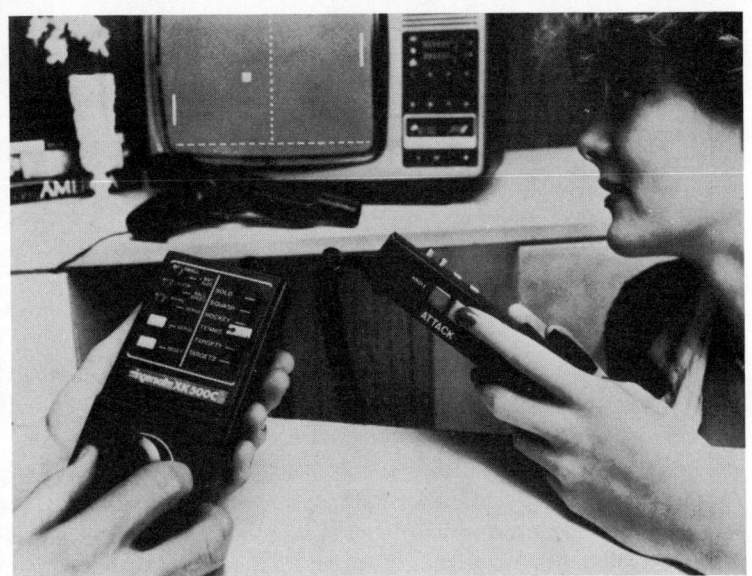

Fig. 11.3　Ingersoll XK500C with six basic games including shooting

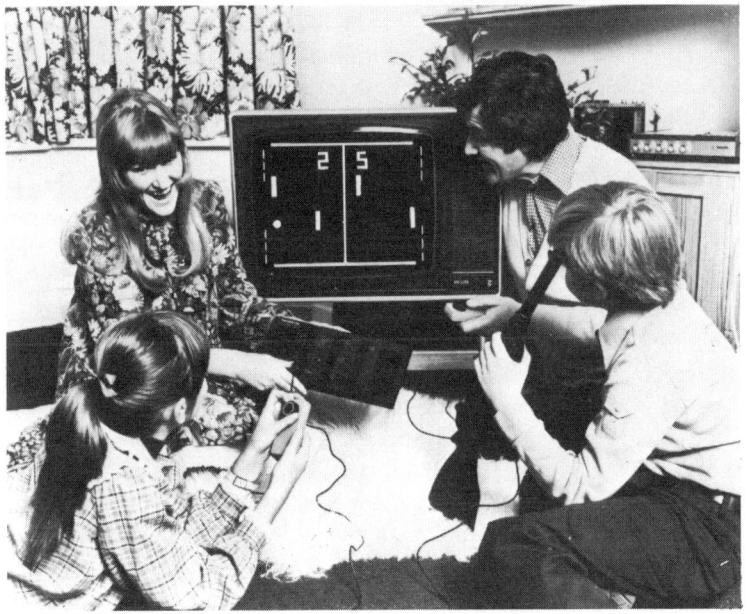

Fig. 11.4　Videomaster Colourscore II with six basic games including rifle

Semi-programmable games

Although the previously mentioned games generally offer good value for money, one can not play tennis and such simple bat and paddle games for ever without becoming rather bored. Although perhaps a veteran, I personally exhausted my patience with simple games over three years ago, and have hardly touched one since (except for review purposes of course). Today, I am rather more interested in the more complex types of games such as chess, blackjack, backgammon, racing and complex battles, most of which hold the concentration rather more than tennis. Again, with all the previously mentioned games, the game selection is somewhat limited, most commonly six, but occasionally ten. With each such dedicated game, there are a number of expensive components that are duplicated, such as power supply, modulator (a micro-miniature TV transmitter that allows the game to be 'picked-up' by the television receiver), hand controllers and cabinet. The electronics for the game are generally contained in a single integrated circuit containing a miniature chip of silicon which generates the game pictures and sound. With this in mind, a new generation of 'semi-programmable' games (only semi- for reasons that will become apparent later), are being produced by a number of different manufacturers, most of which bear more than a passing resemblance to each other in the games and facilities offered although cosmetics identify different brands.

Teleng Ltd, a company within the Telefusion Group, is one company marketing a semi-programmable game, the Colourstars (Fig. 11.5). Designed and built in Britain, the Colourstars accepts plug-in game cartridges enabling additional games to be purchased at a later date. The colour game comes complete with Cartridge No. 1, which offers 10 games including solo squash, gridball, solo basketball, target, solo target, squash, hockey, tennis, basketball and football, all with joystick operation for two motion operation, where required. Cartridge No. 2 Stunt Rider (with accelerator control) gives jumping buses, drag racing, motorcross, etc. for £19·95. Cartridge No. 3 offers road racing, £12·95, while Cartridge No. 5 is Wipe Out, £14·95. Three other cartridges are planned for introduction during 1979, including tank battles, submarine and spacewar. Colourstars operates off either batteries or a supplied mains power adaptor, which is one of the very few available meeting full BS415 safety requirements.

Other companies offering similar semi-programmable games include Radofin with the Programmable Tele-Sports III which also

Fig. 11.5 Teleng Colourstars semi-programmable game

offers seven different cartridges, Optim 600 (Fig. 11.6) with eight cartridges, Novex (from Optimisation) with seven cartridges, Videomaster Colour Cartridge with seven cartridges and usual high quality Videomaster aesthetic design. Finally, Ingersoll distribute the semi-programmable Strike Command (Fig. 11.7) game which sells for £39·95 and includes the usual seven cartridges with road racer, wipe out, submarine and shoot out at £9·95 while stunt cycle and tank battle are £12·95.

Fully programmable games

Semi-programmable games are essentially a compromise between dedicated and fully programmable games. In the first case, the cost of 'tooling up' to manufacture each integrated circuit is tens of thousands of pounds, so they only become economic if vast numbers are manufactured. This was true for the principal dedicated bat and paddle integrated circuit manufactured by General Instruments and the same technique is used for the semi-programmable game cartridges, so the cost of developing new cartridges is inherently very high, and more importantly, the process takes several months. A single

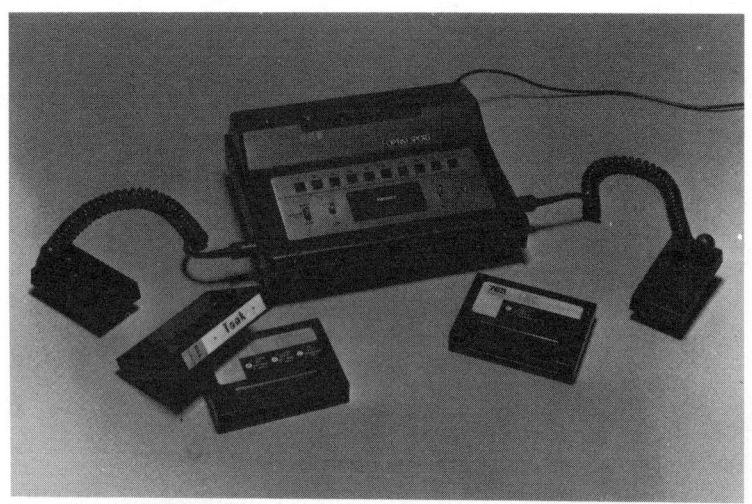

Fig. 11.6   Optim Sport semi-programmable game

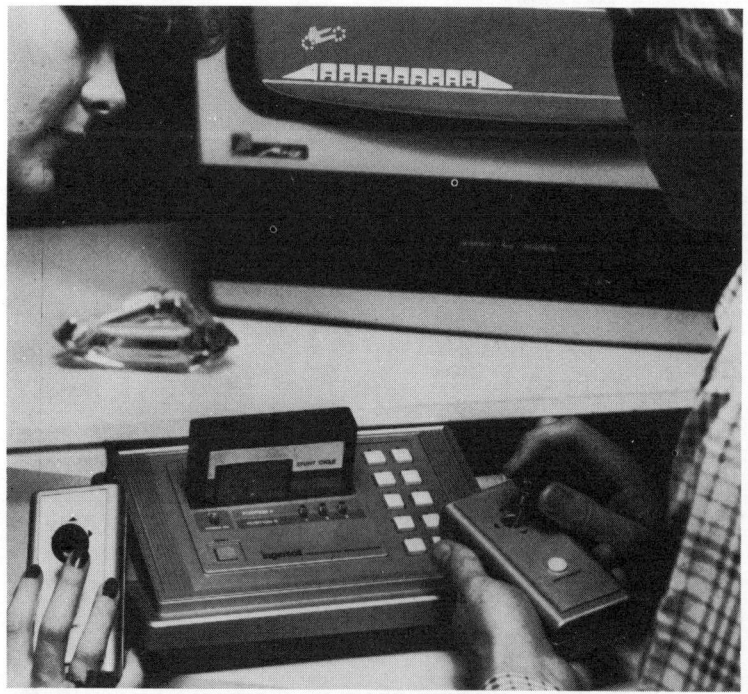

Fig. 11.7   Ingersoll Strike Command semi-programmable game

mistake can mean months of lost production time (as has happened with the tank battle cartridge).

Programmable games operate on a totally different principal – that of a micro computer. In fact, each and every game discussed from now on in this feature, uses as its central electronics element, an off-the-shelf microprocessor. Many of the games from different manufactures actually use identical components and it is the form of programming provided that determines the actual game to be played.

A microprocessor is, on its own, totally useless. It is designed to interpret instructions provided by a computer programme which in the movies is typically seen recorded on a large reel of computer tape, but in practice is now held in yet another silicon chip integrated circuit this time an electronic memory. By mounting this memory in a cartridge, different programmes can be plugged into the game allowing the microprocessor to play different games. The economics of such full programmable games are totally different to semi-programmable games. Almost all the components are off-the-shelf and manufactured in large quantities for many different applications making unit cost very economic. The different memory integrated circuits are produced by a process not unsimilar to photocopying a document – in other, information from a master is photographically copied onto 'blanks', resulting in many identical copies. This procedure is very much cheaper and more rapid than the techniques used for manufacturing the semi-programmable cartridges, and enables much wider ranges of games to be produced, at a potentially more economic price.

Since fully programmable games are based on microprocessors and related technologies it was obvious that semiconductor manufacturer's would be first in marketing the expensive and complex programmable games. Fairchild were first with the Video Entertainment System in the USA in early 1977, and is being marketed in Britain as the Grandstand Video Entertainment Computer (Fig. 11.8) by Adam Imports. Using a Fairchild microprocessor, the unit uses plug-in cartridges containing semiconductor memory (ROM) that may be programmed for a wide variety of games – 24 at the latest count. Games include noughts and crosses, shooting gallery, desert fox, blackjack, spitfire, space war, math quiz, mind reader, drag strip, maze, cat and mouse, backgammon, baseball, torpedo alley, robot war, sonar search, memory match, dodge it, pinball challenge, hangman, checkers, video whizball, bowling, casino poker, space odyssey and American pro football – the last three of which require an additional keyboard.

Fig. 11.8 Grandstand programmable Video Entertainment Computer (from Fairchild)

The normal hand controllers are rather unusual in that the single knob provides eight different motions; push forward, pull back, pull left or right, twist left or right, pull up and push down. Price of the Grandstand Video Entertainment Computer is £119·95 while cartridges cost around £12 each.

Atari (a Warner Communications Company) has been the leading supplier of electronic arcade games since the early seventies and has marketed a wide range of dedicated games in the USA. Atari (the name stems from a Japanese game) has developed a programmable game called the Video Computer System (Fig. 11.9) which is being distributed in Britain by Ingersoll. It is more versatile than the Fairchild game since a variety of hand controllers are available – joystick, keyboard, paddle and steering – for different game cartridges. Again it uses plug-in cartridges containing generally four to six basic games, but many with variations enabling some cartridges to contain up to 50 different games. In fact over 192 different games are available for the Atari Video Computer Systems, the various cartridges available being video olympics, surround, blackjack, basic math, air-sea battle, space war, home run, outlaw, breakout, hunt and score, American football, and basketball. A chess cartridge is also available for about £50, the game itself being currently £169·95 although a price reduction is imminent.

Philips is now manufacturing its Videopac Computer G7000 (Fig. 11.10) in Europe, although the unit was initially developed by a Philips subsidiary in America, Magnavox. Most unusual feature is a full alphanumeric (letters and numbers) keyboard with additional mathematical symbols which allow a wide variety of games to be played, including word games. This facility also potentially allows users to enter their own games into the Videopac Computer using computer language. Games currently available include race, spin out, cryptogram, baseball, pairs, lunar landing, basketball, American football, air-sea war, battle, mathematician and echo (a very interesting game similar to Simon, see later); price of the Philips G7000 is around £150, with cartridges at less than £10.

One rather different entry in the fully programmable game field is the Optim Majestic Model EG1001 (Fig. 11.11) video programmable system since this has a recommended price of only £74·95, considerably cheaper than most others. The Majestic is microprocessor

Fig. 11.9   Atari programmable Video Computer System

Fig. 11.10 Philips Videopac Computer G7000

Fig. 11.11 Optim Majestic Video Programmable System

(Motorola) controlled with two joystick hand controllers and there are currently 12 cartridges available (£12·95 each) with more on the way; targets, grand prix, math quiz, Kung Fu, strategy, chemin-de-fer, blackjack, codemaster, warp-around, checkers, chinense slot machines, and word games. Although the Optim Majestic is obviously good value for money, it should be pointed out that some of the games do not have as many different options as the more expensive games. For instance, blackjack requires two players, whereas most other games provide an alternative single player game – but then you pay your money and take your choice!

Some of the earlier programmable television games have a rather limited display of numbers and symbols since these must be built up individually from discrete squares on the television screen. Since the more displayed squares, the higher the cost of the game, most have rather stylised display for tanks, cars and such like. For instance, the Fairchild programmable game breaks the television screen into 128×64 squares, limiting the size of objects. Some of these games are also limited by the maximum working speed of the microprocessor, which spends substantial amount of its time updating the display memory. Mullard (part of the Philips empire) has developed a new technique for screen display that stores the screen locations of several specially generated objects (such as tanks and planes), which are then recalled specifically for display when required. This allows higher definition to be achieved releasing the microprocessor for more complicated games and sequences.

These Mullard 'chips' are currently used in two fully programmable games – the Interton VC4000 (Fig. 11.12) video computer and Teleng Vision Computer Centre (VCC). Both feature versatile hand controllers which provide a keypad with 12 buttons in addition to the usual joystick. The various game cartridges have cut outs that may be added to the keypad to identify certain buttons as having specific functions. For instance, there are separately identified buttons for hit, stick bet etc. in blackjack, whereas many of the other makes of programmable game use various motions of the joystick to perform the same purpose. Cartridges available include air and tank battles, paddle games (ball sports), equestrian sports, winter sports, air and sea battles, car racing, mathematics, backgammon, blackjack, motorcross, shooting, chess, and intelligence. The Interton VC4000 costs currently £129·95, with cartridges at £12, while the Teleng Vision Computer Centre is around £89.

Fig. 11.12 Interton VC4000 Video Computer

Chess and backgammon game

One age old game attracting the interest of several electronic game manufacturers is chess. For a decade now, it has been the challenge of computer programmers to develop infallible programmes allowing computers to play chess against experts. Many such programmes have been produced, but most needed enormously expensive computers until Fidelity Electronics in Chicago developed the Chess Challenger (Fig. 11.13) in 1976. With this, one plays a standard game of chess against the computer, entering moves on a simple keyboard, and with the computer displaying its moves on indicators – the actual moves are made on a built-in chess board. Various versions of Chess Challenger are available which started with a simple single level unit (which I have not beaten in two years of competition) to the present range which includes the Chess Challenger 10 which is housed in a solid walnut case and offers 10 different levels of play between five seconds response time for beginner level, through advanced (1·2 minutes) to postal chess (24 hours). Computing power is limited by the microprocessor, and so where high levels of play are required, the microprocessor takes a considerable time to scan all the potential moves, counter moves and counter counter moves, and so on. Obviously, the number of potential moves can run into millions, and the longer the microprocessor has to scan these, the higher level of play. Illegal moves are rejected and the board position may be set at any time to replace pieces just captured, or preset specific positions. The price of the Chess Challenger 10 is £200, but the ultimate game must be the Voice Chess Challenger which speaks every move and capture, and offers all the features of the 10 with additional infinite level of play, and display of the number of moves taken during the game. The Voice Chess Challenger uses 96k bits of ROM memory for the program and 8k bits of RAM for recording potential moves – price is around £250. For those not quite up to these high levels of play (such as myself), there is the Chess Challenger 7 which provides seven levels of play and can even play itself as a tutorial aid, and costs £95.

Boris (Fig. 11.14) from Applied Concepts Inc. is another advanced chess computer, offering facilities similar to that of the Chess Challenger 10 such as programmable positioning (to set and change pieces), handicapping, prevents illegal moves, programmable timer to set 'difficulty' and possibly the best feature – it includes an alphanumeric display that 'talks' and comments on the players performance with such quips as 'I expected that', 'good move', 'have you played before' and 'daring'. Boris uses standard chess notation, and unlike the other games uses symbols actually resembling the chess pieces rather than simple numbers – price is about £199.

Fig. 11.13   First Fidelity Chess Challenger

Fig. 11.14 Applied Concepts Boris chess computer

The only British company that has developed a chess game is Videomaster with the Chess Champion. This plays with six different levels which can be changed during the game, and positions may be programmed at any stage of the game. The Videomaster Chess Champion costs £89·50, but unlike the other games, does not include a chess board or pieces.

Atari and Interton are introducing chess playing cartridges for their programmable games and these provide 10 levels of play, and allow either the player against the computer, or two players against each other, using the television for display only. A rather different and absorbing game is Videomaster Star Chess (Fig. 11.15). Here, the basic rules of chess, a game of skill, have been somewhat amended to provide an additional factor of luck. For instance, rather than taking pieces, one can fire missiles instead, each piece having a certain number of missiles and defence shields allowing it to sustain say four attacks before finally disappearing from the television screen into oblivion. A two player game (against each other), Star Chess also allows pieces to be 'warped into hyper space' from where they return a random number of moves later, randomly on the chess board (often on an already occupied square, rather unfortunately). Throughout the game, appropriate explosions, missile whines and pings are produced which also enliven the game. Price is about £60. On the technical side, Star Chess uses a Motorola 6800 microprocessor with 16k bits of ROM and 1k bit of RAM.

Two other board games that have attracted the attention of the electronic game development teams are checkers (draughts) and backgammon. Fidelity Electronics also manufacture the Checker Challenger which is available in 2 and 4 level versions. Features include selection of offence or defence, position memory recall, random computer responses when in identical positions, does not allow illegal moves – the games come with a built-in checker board, and cost £59·95 and £99·95 for the two and four level versions, respectively.

Fig. 11.15 Videomaster Star Chess

Other new games from Fidelity Electronics are the Gammon Challenger, and Bridge Challenger, which can play one, two or three hands depending upon the number of players, and use an optical scanner which reads the cards and can play American style, ACOL and various European bidding systems. They use defensive play, finesse and squeezes, and include 250k bits of ROM (a staggering amount of memory and more than some small computers) with about 8k bits of RAM.

There are two different approaches to electronic backgammon games; self-contained with internal board, and calculator style units which are used with a normal backgammon set. The Gammon Master II from Tryom Inc. (distributed in Britain by Actiongable Ltd) includes a keypad for entering moves, indicators showing moves, dice displays and illegal move recognition. It includes doubling and will play evenly with other experts, and will beat intermediate players and novices most of the time. Gammon Master II uses a Motorola 6800 microprocessor with 48k bits of ROM memory, and 3k bits of RAM; price is about £139. Omar I from Tryom uses the same electronics as Gammon Master II, but contained within a hand held unit. Although it is available separately, when packed with a portable backgammon set the cost is around £100. For expert players there is the Omar II which plays in both modern and classical styles for about £180, while Omar III is similar with a large leather backgammon board at about £200.

Finally, for those still playing conventional games, there are a number of electronic dice on the market (not including the one I built over 10 years ago). Typical of these is the Electronic Easy Dice includes two dice displays and merely require the pressing of a button to display two random numbers displayed in normal dice formations; price £7·25.

Electronic games

Apart from the chess and backgammon, the previously mentioned games have been displayed on a television receiver. While this is usually convenient in the home (but often producing conflict between kids and parents unless a 'two set family'), on holiday, and even in the car, games that require no television display are rapidly becoming available.

Mattel currently produce a range of three hand-held, battery operated games that look rather like electronic calculators at first glance

(Fig. 11.16). Auto Race features a race track with three separate lanes along which collision cars continually move toward the player's car, the trick being to move a guidance control keeping the player's car in a lane away from the oncoming cars. Being rather intelligent, the collision cars always come down the lane your own car is in, requiring constant lane jumping to avoid being hit. Unless hit, the player's car gradually works its way up the track (displayed on light emitting diodes or LED) until it reaches the top. The game is over when four such laps have been completed, the time taken then being displayed. Four gears enable car speed to be increased. The second game is Battlestar Galactica (based on the television serial of the same name), and involves firing laser torpedoes at oncoming 'Cylon Raiders', thus preventing them destroying your Battlestar! Again, the raiders come down three different 'tracks', often changing at the last moment, and one has to guide torpedoes onto the relevant track for a hit. The further from the Battlestar the hit, the higher the gained score which is again displayed. Both games have realistic sound effects, and cost £14·95. The third game is Soccer and involves racing down a football field and avoiding oncoming tackles. Price is slightly higher at £19·95.

Mattel introduced a new range of hand-held games in the USA in early 1979, and these will probably reach Britain in late 1979 or 1980. The new range includes Hockey, Sub Chase, Armor Battle, Baseball, Basketball, and Football (American). All operate on the same principle as the earlier three, but most include a pitch or field of some form. Widening its range slightly, Mattel has also introduced Brain Baffler which features a full alphabetic keyboard for entering words, Horoscope Computer which provides astrological predictions (but only based on birth sign, not date), and finally the extremely versatile Intellivision System that offers a compromise between a home computer and a fully programmable television game with a proper alphanumeric keyboard, hand controller, with overlays for different games, and audio cassette programming. The games all provide much improved display with wide use of colour, graphics and characters, over the present generation of games available.

A range of hand-held electronic games is being introduced by Adam Imports, under the Grandstand title (manufactured by Conic) and these include Electronic Basketball (£14·95), Solitaire, Soccer for two players, and 4-in which includes a calculator, blackjack, road race and mind reader at £19·95. Coleco (distributed by Spectrum Marketing) has also introduced a range of hand-held games including UFO

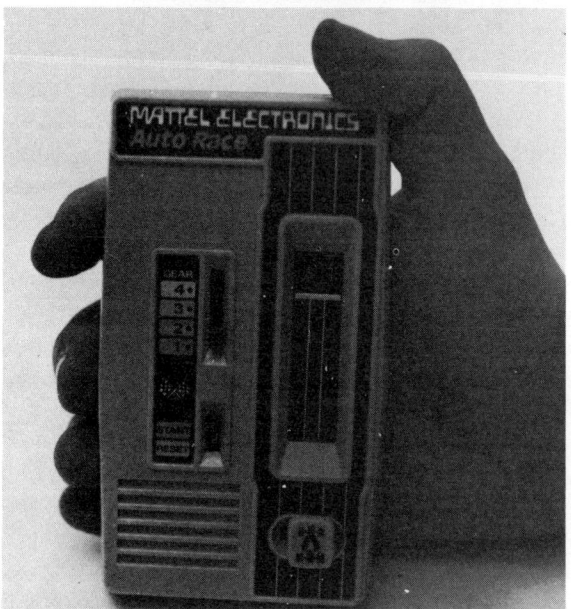

Fig. 11.16   Mattel Battlestar Galactica and Auto Race hand-held games

Fig. 11.17   Coleco Amaze-A-Tron electronic game

Master Blaster Station (£21·95) which requires UFOs to be blasted by missiles (similar to Battleship Galactica but rather larger, with real flying saucers), Digit's code breaker (£13·95) with two levels of skill, Zap! (£10·95) computerised missile game for two players, Lil Genius (£6·45) for teaching simple maths to children aged five and upwards, and Amaze-A-Tron (£17·95) which offers one or two players a choice of eight maze games with sound effects when a wrong turn is made, Fig. 11.17.

One of the large electronics companies in Hong Kong (where the vast majority of television and electronic games originate) is Micro Electronics Ltd. Although the company has launched an electronic baseball game in the USA, this is not expected to be distributed in Britain, although a golf game and football are planned. Incidentally, Micro Electronics manufacture the earlier mentioned Mattel games which use integrated circuits from Rockwell.

Invicta's Mastermind game has become one of the all time best sellers, and naturally there are several electronic games based on the principle of guessing a three, four or five digit number. Unfortunately Invicta has the name Mastermind copyrighted, so these games are distributed under a variety of different names including Logic 5 (Fig. 11.18) (Milton Bradley), Memoquiz (Polymark Ltd), Digits (Coleco), Mind Reader (Grandstand), Logic (Philips), and Codemaster (Optim Majestic). Although the latter three are programmable television games, the others are hand-held electronic types. Invicta's own Electronic Master Mind (Fig. 11.19), is pocket size and uses a small numeric keypad with calculator style digital displays and offers nine different games, based on discovering a number set either by the computer or by another person. Three, four or five digit numbers can be set depending upon the required level of complexity, price being £14·95. Logic 5 is again based on a numeric keypad but instead uses two columns of lights to indicate the numbers of correct digits or correct positions. Again, it plays three, four or five digit games, and price is £17·95.

Sea warfare

Sea warfare has also come under the scrutiny of the electronic game manufacturers, although these are really electronics applied to moulded plastic board games. Milton Bradley has Computer Battleships, Waddingtons House of Games Ltd (same group as Videomaster) has Code Name: Sector, while Action Games and Toys Ltd has Electronic Intercept. Computer Battleships is essentially identical to the age old battleships game, but uses plastic models and peg boards to record moves which are also entered into the computer, direct hits being signalled by appropriate explosion noises and flashing. During play, sonar continuously pings away, while the winner is greeted with 'whoop, whoop, whoop'. Mainly a game for the lower end of its intended 10-to-adult age range. Code Name: Sector is somewhat different since the aim is to trace and destroy of a submarine under control of the computer. A large plotting grid is provided with various coloured crayons which allow the movements of four attacking ships to be traced during the chase to locate the hidden submarine. A 'combat information centre' allows navigation and sonar control to be entered into the computer which indicates directions on a compass and numeric displays. Although a rather complex game to understand and learn, nevertheless very enthralling requiring considerable

Fig. 11.18　Milton Bradley Logic 5

Fig. 11.19　Invicta Electronic Master Mind

concentration. Computer Battleships costs £32·55, while Code Name: Sector is £27·50. Electronic Intercept is another 'search and destroy' game for two players, again with sound effects of battle. One player hides his pieces, while the other player attempts to locate them – a flashing light on the squares signal success; price is £19·95.

Astrology

Coleco have also produced a rather interesting astrology computer called Zodiac (Fig. 11.20). At £25·95, this comes complete with a very comprehensive manual written by Sydney Omarr, and enables full astrological charts to be computed within a few minutes, and full interpretations are also provided in the manual. Zodiac also features a daily preview with appropriate actions for the day, and an advice mode suggesting whether a particular course of action is recommended. In a quick trial, Zodiac computed my astrological chart (based on my time of birth) and this was almost identical to one I had done a few years ago by a professional astrologer. So obviously it does work, although whether one believes in astrology is a totally different matter.

Simon

Finally, Milton Bradley has come up with another electronic game that has been taking the USA by storm, and is set to repeat the performance in Britain. According to press reports, over 600,000 Simon games have already been sold in the USA and that is 600,000 short of demand. Simon is another electronic game, this time modelled on a flying saucer with four coloured lights with matching buttons. Basically, the idea is to echo a sequence of lights (matched with a musical note) which becomes longer with each turn. For instance if Simon says red, red, green, yellow, then the player must then repeat this sequence. If successful, it might become red, red, green, yellow, red, and so on. When the sequence builds up to 15 or more, it becomes quite difficult to repeat, but that is the game – and its catchy. With the tones, it bears parallels to a scene in the film *Close Encounters of the Third Kind* and perhaps it is that inherent simplicity that causes the attraction. Meanwhile, the Philips Videopac Computer also plays a Simon type game, but this time called Echo and using four numbers rather than obvious colours. However, that is the whole point of programmable games – such rages can easily be transposed into television games!

Fig. 11.20   Coleco Zodiac astrology computer

Computer games

Almost all computers at educational establishments and many in commerce have had game software written for them by bored (or just enterprising) programmers, and this expertise has resulted in many games being produced for the home computers that are being introduced by companies such as Commodore Business Machines (PET). Off-the-shelf games include chess, civil war (American style), stocks (playing the markets), policy (running a country's economy) and of course space wars and star trek. The editor of a British electronics magazine has even written a computer program detailing whether a new magazine publishing venture would be successful or not. This is of course where home computers score over other types of video games – you can compile your own games. Unfortunately, home computers still cost several hundred pounds, and apart from playing games,

really require additional peripherals such as computer discs, tele-
types, and printers to utilise them fully. Prices will doubtless fall over
the next few years.

Summary – and the future

It is inevitable that by the time that you read this, developments will
have produced various new games, both television and electronic.
Until now, most of the fully programmable games being marketed
have been developed and manufactured in Europe and the USA, but
the Far East is bound to catch up soon, just as it did three years ago
when the first British bat and paddle games were swamped by imports
from the Far East. Just a word of warning; quality control in the Far
East still leaves something to be desired despite the valiant efforts of
frustrated British importers, so when purchasing an electronic or television game, it is best to ensure that the brand is well known and that
the retailer the game is purchased from will be able to change it, or
return your money (as the law requires) if faults occur during the first
year. There are still a number of 'cowboys' around only interested in
making quick profit from importing a few thousand games from the
Far East, selling that rapidly and then disappearing ignoring their
after sales service responsibilities. To the best of my knowledge, all the
companies mentioned in this chapter are beyond reproach and will
willingly repair or replace any faulty games.

With the existing range of programmable games, electronic memory
is used to store the various different game programs and although this
is reducing in cost very considerably, it is still relatively expensive for
some of the more complex games. With this in mind, EMI and
General Instruments in Britain have developed a new type of programmable game using instead a simple audio compact cassette – this
should enable new games to be sold for only a few pounds, rather than
the present £10 up. Games of this type are expected to be introduced
for the Christmas market in 1979.

Electronic games will also become rather more prominent as the
cost of the ubiquitous microprocessor comes tumbling down – it will
only be lack of ideas suitable for hand-held games that will restrain
the market, and potentially cost should drop to below £10 as production runs increase. Prices are however unlikely to fall as significantly
as did those of calculators, since large sums of money will still be required for research and development, producing new models for each
year's Christmas market. Of course, television games are restricted to

use in the home, and they obviously mean that normal television programmes can not be watched by other members of the family (apart from those families lucky enough to have two sets). However, electronic games can be used virtually anywhere without restriction, so can keep kids quiet in the back of the car, or on holiday during rainy weather. They do all tend to gobble up batteries rather fast (usually PP3s), and not all games have a socket allowing a mains adaptor to be plugged-in. For those games lacking this vital facility, a mains adaptor with battery type terminal could be plugged directly onto the battery clips, although this would probably not allow the battery compartment lid to be replaced.

Television games only really started in 1973, and yet since then we have progressed from the most basic bat and paddle games, through multiplayer games, racing, battles, blackjack and chess. The future should prove just as intriguing.

12   Home Computers

Richard Elen

The word 'computer' conjures up an image of a roomful of whirring machinery, bright lights, rotating tape-spools and not a person in sight – or at least it used to. Such massive computer installations still exist, of course, and commercial computing departments still have the reputation for sending you bills for £0·00, payable in seven days or they will cut off the electricity! But computers are not mechanical ogres, or even electronic ones, although Government departments and other organisations sometimes seem bent on making them so, with untold pages of information of doubtful accuracy on as many of the population as possible, which without a Freedom of Information Act we can never check. In fact, a computer need not take up a room either. Indeed, a modern computer with more 'power' and calculating ability than a roomful of 1960 computer equipment can sit on your desk and can cost as little as a cheap colour TV.

Microminiaturisation in electronics has brought about this revolution. When electronic computers were first built, some of them were simply huge: the Ferranti Star Mark 1 machine, for example, which was used to calculate the equations for the first atomic bomb, contained literally thousands of thermionic valves and required a large cooling system. A computer of the early 1960s, with the advent of transistors, was down to a few six-feet cabinets. Then the development of integrated circuits (ICs) made it possible to cram the equivalent of hundreds of transistors into a small 'chip' of silicon, considerably smaller than your fingertip. It is just such a chip which is at the heart of a modern microcomputer. Each silicon chip is mounted in a plastic package with as many as 60 pins; peek inside any modern computer and you will see rows of them. In a microcomputer, the biggest IC noticeable is the *Microprocessor*. This, in computer jargon, forms the

CPU – Central Processing Unit – the 'brain' of the computer which processes all the information and does all the calculations. The other ICs perform separate but related functions and the whole is 'The Computer'.

Just what is a computer? Well, it is simply a combination of a storage system and a calculator. The storage is basically electronic record keeping – information is stored by the computer for use in a number of ways. In the computer itself you will find a number of chips of 'memory'. Some of these are known as 'Read Only Memories' (ROM) and are used to store the computer's operating systems and procedures that enable the computer to process the information you give it, and organise the computer's inner workings. Other memory chips are known as Random Access Memories (RAM) – these are used to store your programs when they're being run, and the data you give the machine to work on, plus the data it produces to work with. Still more information, both programs and data, are stored in 'peripheral' memory units, like magnetic discs and tapes.

The computer can also handle complex mathematical calculations at very high speed, and this combination of storage and calculating ability makes the computer a very versatile tool. However complex computers may be, without programming they are just plain stupid, because they need programs (sets of instructions) in order to tell them what to do. These programs are known generically as 'software', distinguishing them from 'hardware' the equipment that constitutes the computer; and 'firmware', the special operating systems and languages that convert your instructions into a form the computer can understand. We will come back to programming and programming languages later; first let us look at the rest of the machine.

We've seen how the CPU does all the calculation, and how the memory stores your programs in RAM. In fact, the 'firmware' we mentioned above, whilst it is stored in a 'memory' of sorts (the ROM which stores information permanently), is resident in that part of the computer which is known as the control unit. This allows the computer to execute the right instructions at the right times. The remaining two parts of the computer are the input unit and the output unit. The input unit enables information to enter the machine in the form of data, control signals and software. Basically, it forms the *interface* between the CPU and peripheral devices used for inputting information, for example a keyboard, tape cassette or disc memory. Likewise, the output unit interfaces (or 'connects' if you don't like the jargon!)

the CPU to the various devices used to display and record information, for example a TV screen, cassette, magnetic disc, or a printer (Fig. 12.1).

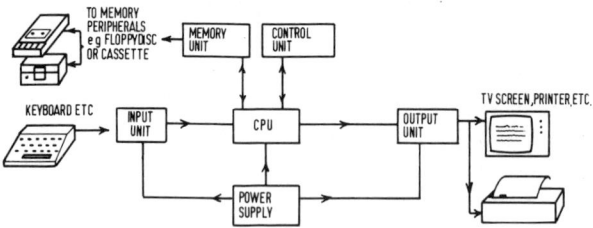

Fig. 12.1 Schematic arrangement of a basic home computer system

A look at software

Computers are fairly complex pieces of electronics, but that does not mean to say that you have to know all about them to benefit from them. Just as you do not need to be a mechanic to drive a car as you only need to learn how to use the car – similarly, you do not have to be an electronics specialist to 'drive' a computer. In fact, all you need is to know how to 'talk' to the computer to tell it what you want it to do. You do this with the use of a 'programming language'. A language to do is required because the microprocessor itself uses what is known as 'machine code', which is not very intelligible to ordinary people, although computer cognoscenti often program directly in machine code (amongst other things, it is very fast because you are 'talking' to the computer in its own language). We ordinary computer users have to converse with the computer through an interpreter: a special language which is close enough to English for us to understand, yet is sufficiently closely-defined to be unambiguous to the machine. The machine's interpreter converts our special language into machine code.

These 'high level languages' as they're called have been developed over the years often for specific applications, like mathematical or business programming and have names like FORTRAN, COBOL, PASCAL, and others. Most common of all in the modern world of desk top personal computing is a language called BASIC; and 'basic' it really is. It is easy to learn, and simple to understand – in many ways it is languages like this which have taken computers out of the hands of specialists and put the microprocessor in a position where we can all experience the benefits.

Of course, the next question, before we actually take a look at choosing machines, is what exactly are these benefits? What can we use a personal computer for? Well, as we have seen, a computer can perform complex mathematical calculations, and it can keep records. It can also be used to control other equipment, from complicated automatic machine tools (even entire factories), down to simply switching the lights on and off. It can play games with you or your children, and can even play music for you and help you compose your own (like the microprocessor based Micro-Composer produced by one musical instrument manufacturer). In fact, the microprocessor is becoming very important in the production of modern music. A personal computer can also be used by small businesses for payroll, stock control and accounting, or to calculate taxation for the self-employed, mortgages and budgets for the householder. Children can learn (as can adults) a vast number of skills with computer aided education programs, which can respond individually to the student's needs.

Yet the main reason for the popularity of personal computers is sheer fun. It is fun to write programs, play games (Fig. 12.2) and so on. You do not need to be a professional programmer. BASIC is easy to learn, and for the more complex programming requirements, most of the better personal computers are well backed-up by comprehensive program libraries for every conceivable application. Householders in the USA (and no doubt elsewhere) are using computers with voice-recognition, so they can switch on the lights or open the garage door merely by speaking to the computer, saying no more than 'Computer, switch on the lights'!

Most personal computers include a typewriter style keyboard for programming and 'talking to' the machine, and a connection to a cassette recorder to load purchased program cassettes or to save your own programs for later use. In fact, very often all you need is a TV set to display the computer's output, and some machines can even draw pictures and diagrams in colour! Personal computing is a fascinating world, and one which will provide years of pleasure – and usefulness in the home, business or education. Books, like Alcock's *Illustrating BASIC* (Cambridge University Press) can teach you programming with ease and a little perseverance, and various monthly publications will keep you informed on new developments and interesting programs and applications which you can try at home.

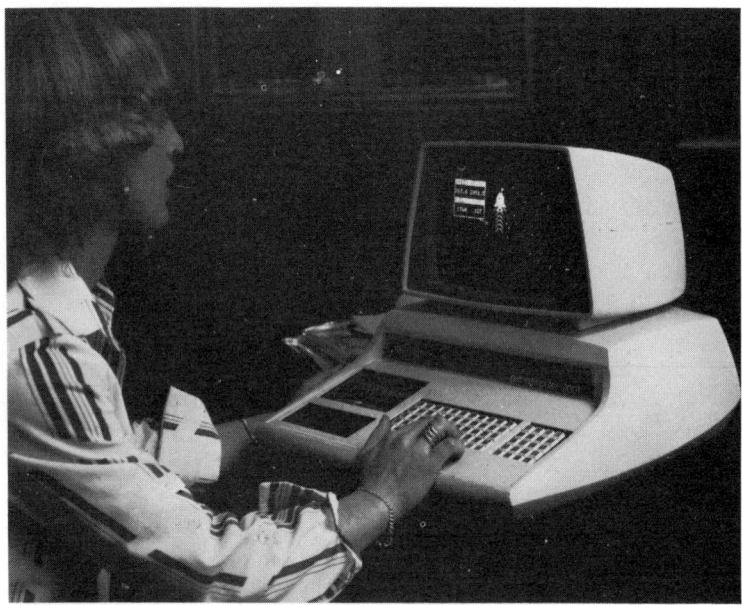

Fig. 12.2 The popular Commodore Business Machines PET computer which has a built-in cassette recorder and offers options for floppy disc and printer. There are many PET computer clubs where members exchange programs, and numerous other programs are readily available

A personal computer is as simple to set up as a hi-fi system, if not simpler, and machines are on the market which range from simple machines which will only cope with prerecorded 'bought in' programs right up to sophisticated programmable machines like the Apple II and ITT 2020 (Fig. 12.3), which offer graphics, colour, sound and many other useful (and amazing) facilities and Atari 800 (Fig. 12.4). To determine the best machine for yourself, just work out what you need, how much you can afford, and so on. If you know what your applications for the machine will be, you can evaluate the systems which are on the market – and these are many. But make sure that you allow room for both you and your computer to grow!

Questions to ask
When choosing a personal computer, you should look at these general areas:
1 The manufacturer Just as when you buy a hi-fi system, it is

worth looking for reputable manufacturers and dealers. In an industry which is growing as fast as this, there are regrettably several 'fly-by-night' operators who are offering badly-designed gear with little or no backup for service or software. Check magazines and people you know who have machines; find out just what the manufacturer offers. Is there a guarantee? Is it suitably long? Is the machine sturdy and well-built? Is it quickly repaired if it goes wrong? Does it have to go back to a distant factory or can it be repaired locally by your dealer? Are dealers happy with the service they get from the manufacturer? All these and other questions should be satisfactorily answered.

2 Is it easy to use? The microprocessor chip at the heart of the computer will be made by one of a handful of reputable manufacturers, and is unlikely to be a problem. This is not necessarily the case, however, with the rest of the machine! Is it easy to plug in and get running? Is it easy to program? Does the keyboard have a good, positive 'feel', and does it always do what it should, or does it tend to produce multiple or odd characters when it shouldn't? Will the machine help you learn about computers as you use it? Will you find it limiting? Does it perform the right tasks for your application? Has it enough memory for what you want to do? Can you play games on

Fig. 12.3 Apple computer (ITT model shown here) which provides a colour picture on a separate receiver, and uses cassette recorder and many other peripherals

Fig. 12.4   Atari 800 computer with printer

it? The questions to ask yourself about a computer will depend on what exactly you want it to do – be sure you know!

3 Will it grow?   Your needs are bound to expand as you become more experienced. Particularly, you may find yourself wanting more memory to store and run longer, more complex programs. So has the computer enough RAM? RAM storage is measured in bits or kilobytes, and you should check that there is enough, or that you will be able to add more memory boards. A machine which offers no more than a basic '3k RAM' will be fine for short programs, but you may eventually need as much as 32k or even 48k. Will the machine allow you to extend? Is extra memory easily fitted? Then, can you store programs easily for future use? Most machines offer cassette storage, and you can load and save programs with a simple cassette recorder. This is cheap, but slow. If you are going to be running and storing a lot of programs, perhaps in a small business application, floppy discs can offer large amounts of storage on magnetic discs. Programs can be quickly and simply saved or recalled, often in as little as a few seconds, whereas a complex program might take many minutes to load from a cassette. They are expensive, floppy disc drives, but you may well need them. Will your machine let you? Then, can you add a printer, so you can obtain 'hard copy' on paper for accounting and the like? Can you hook up accessories, like voice control or external switching, so the computer can control other devices easily, like burglar alarms? Expandability is very important, because as time goes by, your needs will no doubt change.

4 Sound and graphics   Most personal computers give you the output via a TV set, either a standard TV or a specially-modified set with a video input (this latter gives better picture quality, but is a little more expensive to implement) in black and white. Some of the better models provide colour and graphics as well. Graphics enable you to draw graphs, diagrams and the like, and colour can make the computer output more understandable and more easily read by other people. Of course, an accessible readout is primarily the result of good programming, but colour and graphics can help a great deal. Do you need them? Some computers also produce sounds, not just for creating music, but also for giving you an audible indication of errors, and making games more realistic by producing 'bips' when you hit the ball, or explosions when you destroy an enemy starship (!). Sounds can also be used for alarms. Consider these possibilities if you think you will need them – awareness of your needs once again.

5 Peripherals   Peripherals are things you hook up to the computer to increase its usefulness. Things like printers, floppy discs, and the like, are they easily connected? Or do you need to modify the machine expensively? Do you have to drastically alter your programs to suit, or is the adjustment simple and easily written? Are there a large number of peripherals available, or are they 'promised for the future'? They may never arrive. A system with a good choice of peripherals today will ensure that you will be able to get what you want when the time comes. And when you do need your peripherals, you will probably find that the situation has improved still further.

6 Manuals and instructions   Most computers come with a book of instructions, telling you how to set up the system. This should be clear and easy for *you personally* to understand. There should also be much, much more. Look for a book on the programming language the machine uses, because whilst the text-books on a given language are often useful, there are many varieties of BASIC and the other languages you might end up with, and there are quite large variations between one version of BASIC and another. Again, you should be able to understand the book. There should also be a 'reference manual' explaining the machine and perhaps giving you example programs so you can fully utilise the facilities offered that are specific to the machine in question. Sometimes teaching aids are provided in the form of software programs that let the machine introduce itself to you. Run them and see if they indeed do the job! Ask to see all the manuals that go with the machine before you buy, and make sure you can get along with them.

7 Software   Last but not least. When you get your machine, depending on your requirements and your interests, you will want to run not only your own programs but also buy in programs designed for specific applications. Some things, like record keeping and accounting, are quite hard to write unless you are already an experienced programmer (and if you are, you will not need to read this!). In addition, why spend ages writing a program when someone else has already done the work and it costs you little to buy? So ask about the 'software backup' available. Is there extensive program library? Can other languages be supplied? What commands and instructions are available in BASIC or other languages? Can languages be supplied on cassette, plug-in ROM cards or cartridges, or floppy disc?

I hope these tips will be useful if you come to purchase your own personal computer – and I certainly hope you will do so. It is a fascinating subject and one which will give pleasure and usefulness for years to come. Enjoy it, and have fun!

13 Electronic Watches and Calculators

Chris Webb

One of the biggest impacts made by microcircuitry on consumer electronics has been on the digital watch and pocket calculator market. Within the last five years, the public has been confronted with two highly complex yet relatively cheap packages which offer accuracy in time-keeping and mathematics, probably far in excess of most people's everyday needs. Ten years ago calculators were heavy, clumsy machines which were only employed in accounts departments as their abilities were limited strictly to simple mathematics. Anyone who wished to solve day-to-day problems had to use slide rules, scraps of paper, or their fingers – yes it really was as recent as that and yet the calculator has made such an impression on our society that life without it seems quite incredible.

About one in four adults now use a calculator, and prices range, according to complexity, from around £4 to anything up to several hundred pounds. Such is the impact on our ability to now solve complicated problems almost instantly, that schools are beginning to incorporate the use of them into mathematics classes!

At first glance, a collection of calculators presents a confusing array of very diverse machines. Some of the more advanced ones can perform specialised operations such as computing the tangent of an angle or the power of a number in a single key stroke (Fig. 13.1). Advanced business calculators are capable of solving many of the mathematical problems encountered in the world of commerce. Brokers and statisticians use them to evaluate data, analyse trends and make forecasts. Calculators are also used in such diverse fields as biology, agriculture, aviation and meteorology (Fig. 13.2).

To the lay person though, its a good thing to ask oneself exactly for what purpose one will be using a calculator before purchase. Most of

Fig. 13.1 Commodore (CBM) SR9190R scientific calculator with nine memories and LED display

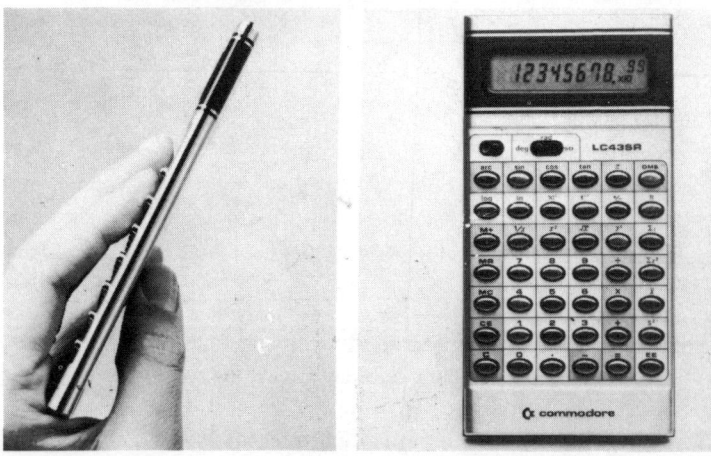

Fig. 13.2 Commodore LC43SR scientific with 8 + 2 LCD display

the simple cheaper calculators offer quite a wide range of abilities including simple addition, subtraction, division, multiplication and usually include a percentage key plus a simple memory for temporarily storing any answers to one problem, thus enabling the user in the meantime to solve others.

But moving onto the more advanced models, one starts dealing with pre-programmed and programmable calculators. Pre-programmed calculators have specialised mathematical operations built into them and there are two types: those which feature scientific operations and those which feature business operations. Programmable calculators have a number of pre-programmed operations but in addition they can 'learn' how to solve complex or repetitious problems in a variety of fields quickly and efficiently.

Let's look at the pre-programmed calculators first. The typical scientific calculator derives logarithms and trigonometric functions far faster than a slide rule. It can calculate angles in either radians or degrees and its ability to extract roots or raise numbers to a power is only limited by the range of its numeric display. Advanced business calculators (Fig. 13.3) are capable of calculating a wide range of problems encountered in the business world and usually include a range of specialised keys for such financial problems as interest rates, payment period, payment amount etc.

Moving onto programmable calculators we start nearing the field of computing. Some of the programmable calculators with their abilities in printing (Fig. 13.4) and display techniques make the capability gap between calculator and computer very small indeed.

When IBM introduced its 650 computer back in the middle fifties, it occupied 45 square feet of floor space, weighed nearly three tons and required over five tons of air conditioning, yet a modern hand-held programmable calculator with similar functions requires only one hundred-thousandth of the power and contains the equivalent of the 166,500 transistors which drove the IBM 650. Possibly the only area in which this type of calculator compares less favourably with a mainframe computer is in memory capacity, but this area will no doubt be shortly rectified.

In the last few years programmable calculators have moved up from the simple keyboard programming stage which limited the number of program steps available and also meant that the program was lost when the machine was switched off. Such facilities are now available on relatively low cost machines and at the top end of the market

**Fig. 13.3 Texas Instruments T1-51-111
programmable calculator**

we have seen them augmented firstly by magnetic card programming features and, more recently, by plug in memory modules containing complete sets of standard programs.

This capability means that one can customise a fully programmable calculator for virtually any area with important facts to help make an intelligent, knowledgeable purchase decision for instance. The better machines also incorporate a special set of instruction keys specifically designed for this purpose. These keys permit the user to edit his way through a program one step at a time, to delete errors and correct mistakes.

The usefulness of a programmable calculator is very much enhanced by the existence of a wide range of programs in various disci-

Fig. 13.4 Hewlett-Packard 19C programmable printing calculator

plines. Although the owner can build up his own library, it saves considerable time if he uses those prepared by others. Some of the manufacturers of these calculators provide users with prepared program libraries for a small fee, or programs can be obtained from calculator clubs, universities and various publications.

As far as the future of the calculator market is concerned, the cheaper simpler models will incorporate more and more attachments as the trend already shows. Many incorporate watches (Fig. 13.5), radios, even lighters (Fig. 13.6) and so it is with the digital watch which is also beginning to broaden with more sophisticated functions. Already there are digital wrist watches that also act as a calculator.

The UK digital watch market has grown quite steadily over the

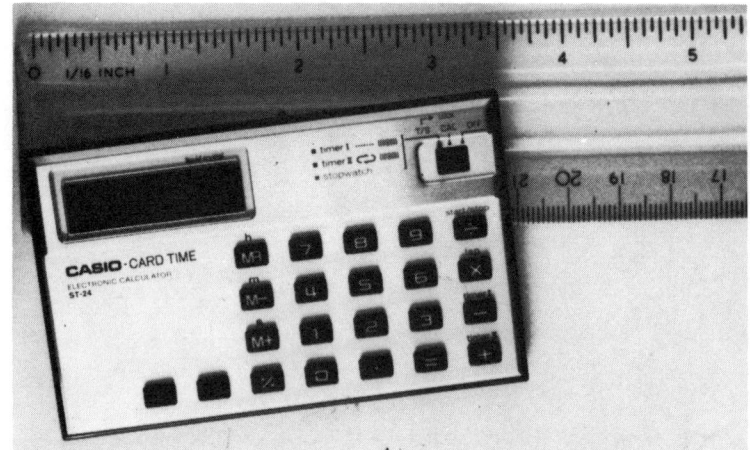

Fig. 13.5   Casio ST-24 Time Card calculator with watch and timers

Fig. 13.6   Casio QL-10 calculator complete with alarm, watch
and cigarette lighter!

past eight years but the real revolution began in 1975 when watches reached a new height of accuracy, and innovation at a lower cost compared with traditional watches. In 1978, digital watches (Fig. 13.7) captured 38% of the watch market and this figure is expected to grow even further.

The would-be purchaser of an electronic watch has the choice of either a quartz analog (Fig. 13.8) which combines the accuracy of modern technology and the old fashioned appearance of a dial with hands, or the digital type (Fig. 13.9) which uses an LCD (liquid cystal display). However as watches incorporate more complex extras such as calendar or calculator functions, it would seem that the digital watch will completely dominate the electronic watch market. The old fashioned LED (light emitting diode) type display will probably fade away as the LCD type of display also becomes better with a range of different colours, etc. The LCD is a passive display which uses the contrast between the surface which either reflects or absorbs direct light. Its main advantages are that it has a very low power requirement which enables continuous display (unlike the LED display

Fig. 13.7   Range of LCD watches from Unik Time

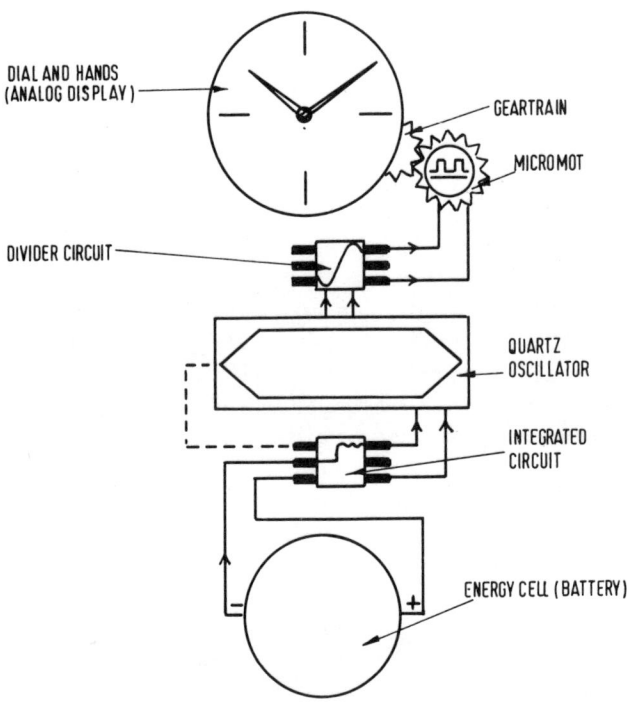

Fig. 13.8 Quartz analog watch module

which has to be switched on for a reading), and for this reason it is expected to become the standard display with minor variations.

The electronic watch really goes back to 1967 when Bulova marketed its Accutron which used a resonator like a tuning fork which vastly improved accuracy over the standard mechanical watches. Then in 1972 the company brought out a watch which incorporated a quartz crystal oscillator as a time base. This basic element, either natural or synthetic, provides the frequency oscillation which is controlled to one oscillation per second by the integrated circuit.

Like the calculator, the prices of electronic watches have plummeted. Within the UK alone, electronic watch sales are expected to grow by about 30% over the next two years. The dominant type of electronic watch is the five and six function type (hours, minutes, seconds, date, month and day of the week), with other extras like alarm and stop watch.

224

As in the higher priced models, it is expected that aesthetic appeal will play a significant role in expanding the cheaper end of the market. One of the chief criticisms levelled at digital watches is that they are so damned ugly. Also, they are still fairly bulky, although time should bring improvements.

A recent development which may affect the complex evolution of the market is an innovation from one of the biggest manufacturers of watches, Texas Instruments, who have introduced a watch with an analog shaped liquid crystal display which, if proved viable, may affect the watch market in the forthcoming decade.

Another criticism made against electronic watches is that the average battery life is not usually more than a year or so. However the advent of the solar cell should rectify this.

There is no doubt that the solid state digital watch is the most accurate method of time keeping available. In addition, the digital display provides a more efficient and precise method of presenting time of day. On a digital device, time is presented instantly and accurately to the fraction of a second, while on the analog device, time is

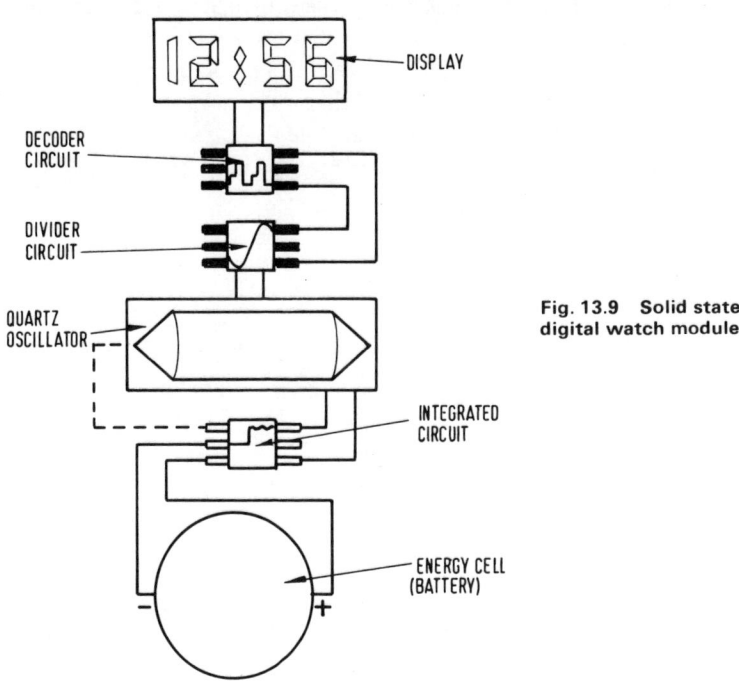

Fig. 13.9 Solid state digital watch module

presented as a relationship of hands to dial marker and no matter how constant it may be in its working, it can never instantly convey the exact time. However because the digital display is not as attractive as the analog, it has proved difficult for the manufacturers to sell their watches to the middle and top sectors of the market. However it would seem almost certain that this reluctance will be overcome as the consumer becomes more familiar with the strengths of digital timepieces and as the manufactures continue to improve appearance.

Last year one of the big names in the calculator and electronic watch business introduced a microcircuit which 'speaks'. Already this chip has been incorporated into a children's spelling game whereby the calculator calls out the letters of the alphabet as each key is pressed and also calls out complete words – though these are pre-programmed. It is interesting to speculate whether these talking chips will ever be incorporated into standard calculators or watches so that everyone will be carrying their own 'speaking clocks'. There is no doubt that as microcircuitry gets even smaller, more and more functions are to be incorporated onto our watch/calculators. Perhaps intricate bio-medical capabilities to check out temperature, pulse rate and blood pressure. A pager, to call us when we are wanted on the phone; hardly improbable when 10 years ago we were all relying on clockwork and slide rules. One thing is certain and that is the market has still a long way to go in producing more intricate and complex wrist packages to enable us to measure and count our increased leisure time.

14   Home Protection and Security

Christopher Chambers

Many of the earlier chapters in this book have been devoted to products that have been, or will be, made possible by the advent of large
scale integrated circuits (LSI) in general and microprocessor technology in particular. Such technology will undoubtedly make a tremendous impact on many industries and occupations in the eighties
by considerably reducing the number of man hours (person hours?)
required to perform a wide variety of tasks. It is therefore fitting that
the technology that will make shorter working weeks possible for large
sections of the population, will also provide them with the means to
occupy their increased leisure hours. Unfortunately, however, it
seems as though one of the prices paid by affluent, consumer orien-
tated societies with rising standards of living is a rapidly rising crime
rate. Even a casual observer cannot escape the constant reminders by
the mass media that nationwide levels of lawlessness have reached
record heights, and a more detailed examination of crime statistics
emphasises the seriousness of the problem in the UK. In 1977, the
total value of cash and property stolen in thefts, robberies and burglaries that were reported to the police was £198 million. Of this
staggering amount only about £31 million was subsequently recovered, so criminals made a tax-free profit of nearly £167 million!
When one considers that these figures do not include losses from em-
bezzlement, fraud and forgery, or theft from retail premises – which
most authorities estimate to exceed £600 million annually, then the
old adage 'crime doesn't pay' is seen to be fallacious in modern society.

Of course, many of us dismiss such grim statistics with the unjustifi-
ably optimistic philosophy that 'it won't happen to me' but the law
of averages says otherwise. In London, the total number of recorded
crimes exceeded half a million in 1977 – that means one every minute

of the day. In the country as a whole, there are nearly 12,000 burglaries every week or about 70 every hour. In more personal terms, this means that one in every 30 homes can expect to be burgled during the year and only four out of every 10 households who suffer this fate can expect to have any stolen property returned. In London the odds of being burgled are one in six, whilst one in 12 homes are burgled *twice* every year! As the crime rate has increased by 50% in the last four years, we can only assume that the situation will get even worse unless a totally different approach to crime prevention is adopted in the future.

Ever since the last war, successive British governments of whatever political party have implied that their policies of social reform together with the general improvements in the standard of living and the welfare services, would collectively contain or even reduce the crime rate. At the same time, the strength of the police forces has been maintained at levels commensurate with these policies – that of the London Metropolitan Force for example has been kept at the 1920 establishment level, and police pay has hardly kept pace with the wages enjoyed by thousands of men and women in much less demanding careers.

The consequences of such policies are clear. The crime rate, which has increased fourfold since the early 1950s, is rising faster every year and the police have totally inadequate resources to deal with the problem. This situation has prompted frank and uncompromising statements from such highly regarded and knowledgeable figures as Sir Robert Mark, formerly commissioner of London Metropolitan Police, who, whilst commenting upon the inability of the state to effectively protect people from burglary and theft, said 'The time has come to explain frankly that each and every citizen must take the primary responsibility for protecting his own property. . . .' But how can we do so? On the one hand we fill our homes with desirable, high value, low volume items that are comparatively easy to steal, whilst on the other hand, every burglar knows that he runs little risk of arrest and punishment when he steals them.

Historically, the classic defences against intruders that were established even before man started coveting his neighbour's goods were, and still are the dual weapons of prevention and detection. The knight or king in his castle *prevented* or deterred the entry of his foes by providing physical defences in the form of moats, drawbridges and high walls. At the same time reliance was placed upon *detection* of intrusion attempts by the deployment of sentries who were posted whenever danger threatened the king in his castle.

In modern times these concepts were, and are, employed by those threatened by criminal intrusion although until comparatively recently, only very rich Englishmen could afford to protect their 'castles' efficiently.

Nowadays, the fortification of our dwellings is restricted to the provision of physical defences in the form of locks, bolts, bars and grilles whilst intrusion attempts may be detected by burglar alarms. Contrary to popular belief, fostered no doubt by the cinema, TV, and written works of fiction, intruder alarms cannot *prevent* unauthorised entry – they can only detect intrusion attempts and generate a warning which, hopefully, will be acted upon. Unfortunately, however, an intruder alarm's ability to reliably detect entry is also less than fiction would have us suppose simply because, in reality, even the most complex systems are unable to discriminate between genuine intruders and many naturally occurring phenomena.

This basic lack of discrimination can be illustrated quite simply. Suppose one attaches a switch of some sort to a door and then arranges things so that when the door opens a bell rings (exactly what happens in thousands of alarm systems). Now one might suppose that this very simple arrangement is reasonably foolproof but just consider for a moment what the bell is actually signalling when it rings. All such a warning can tell us is that the system has reacted to some change of state. Did a burglar come through the door? Was the door catch faulty and a draught blew it open? Did the switch or the connecting wiring fail? We simply don't know and therein lies one of the fundamental problems with all security systems. Even when the equipment is operating reliably, many false alarms can be generated. To put this problem in perspective, one only needs to examine police statistics. In the London Metropolitan area, alarm systems of the kind that give warning to the police either by automatic telephone dialling machines that dial 999, or by means of privately rented post office lines that connect alarm systems to the premises of the installation companies directly, generated 169,834 calls in 1977. Of these, 167,256 were false alarms leaving less than three thousand true alarms. These statistics do not include the false alarms generated by burglar alarms that rely solely upon an audible warning, usually in the form of a bell.

Of course, such figures can not adequately illustrate the tremendous wastage of police man hours caused by attendance at premises when false alarms are generated, nor do they reflect the aggravation of alarm users who are called from their beds to attend their business premises in order to reset their systems. Such a situation prevails in spite of the introduction of a British Standard for alarm systems which all the leading companies undertake to comply with; in spite of the activities of the National Supervisory Council for Intruder Alarms (NSCIA) whose inspectors check for compliance with the British Standard (BS4737:1971 and later revisions); and in spite of a new policy towards false alarms now implemented by many police forces whereby response to alarm calls may be denied when systems generate more than a predetermined number of false alarms in a particular period.

It is perhaps small wonder that, faced with such an unhappy situation, by far and away the largest proportion of the 25,000 or so new alarm systems that are installed every year by NSCIA approved companies are reserved for commercial or industrial premises or for wealthy home owners who have to have systems installed as a condition of insurance cover. In latter years many small alarm companies, and indeed one or two of the larger concerns, have provided systems tailored more for the average domestic residence but the number of installations in the private sector is still minimal. The reasons for this are many and varied. As already mentioned, many people mistakenly believe that burglary will not happen to them. Others are apathetic towards practical preventive measures, mistakenly believing that insurance is a substitute. (How many people realise that they are liable if a rented TV is stolen and should therefore include it in their 'contents' insurance?) Many more are undoubtedly put off by the charges levied by security companies for their products and services whilst some people, although mindful of the need to protect their property, are put off by the experiences of others. Unfortunately, complaints from users that their system is too complicated to operate, or that there were 'wires run all over the house and we had to redecorate every room' are fairly common.

All in all one might be forgiven for thinking that the installation of intruder detection equipment is not worth the time, trouble, inconvenience and cost and that the installation of high security locks and other physical protection is a better proposition.

Fortunately, however, these attitudes may well change in the near future because technology itself holds the key to new generations of alarm systems that are very much more cost effective than their conventional predecessors. Before examining such developments we should first review the components and operation of existing systems.

All alarm systems comprise of five separate, but related elements (Fig. 14.1). The heart of the system is the *control panel* which provides the means for turning the system on and off as required. The second element comprises the *detection devices* which respond to unauthorised entry through protected doors and windows in the building shell or react to movement within the building. These detection devices are linked to the control unit by the *circuit wiring*. When the system is turned on, the control unit constantly monitors a flow of current through the circuit wiring and is thus able to react to the interruption or modification of this current flow that occurs when a detection device operates or when the wiring is accidently or deliberately broken. The fourth element is the special *power supply unit(s)* (PSUs) that delivers any of the other elements of the system with the correct current and voltages they require. Such PSUs invariably incorporate batteries that take over whenever a mains failure occurs. The fifth and final element is the *signalling system* that generates a warning whenever the control unit detects a change-of-state in the protections. As previously mentioned, such warnings can be just audible devices, like bells or sirens, that sound locally or, for premises with a higher burglary risk, the signalling system may transmit a 'silent' warning to a remote location like a police station or an alarm company's central station.

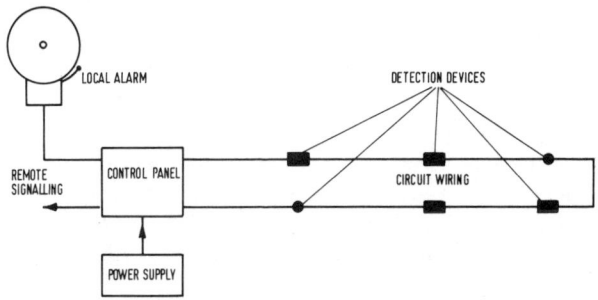

Fig. 14.1   The components of an alarm system

A wide variety of detecting devices have been developed for use in such systems and range from simple magnetic switches for use on doors and windows to very complex electronic motion detectors that sense the movement of intruders once they have gained access to the protected property. Within the confines of such a book it is not possible to describe in detail the function and operation of every type of detecting device but, broadly speaking, they can be categorised into two groups. The first group provides what is known as perimeter protection in that they are used to detect entry through doors, windows (and sometimes walls and ceilings as well) in the building shell. Such devices include magnetic contacts; vibration contacts, both mechanical and electronic; infra-red rays and, in many instances, the circuit wiring as well. In this latter case, the monitored wiring that links the detecting devices to the control unit is used to protect vulnerable doors or panel walls by lacing it back and forth to form a mesh. The wiring is then concealed by cladding the door or wall and the cladding material also serves to protect it from accidental damage. If an intruder attempts to break through the wall or door the circuit loops are broken and the alarm sounds.

An extension of this technique is often employed in industrial and commercial premises to protect glazed openings. The circuit loops are run through metal tubes fixed between timber battens to form a grille which is then securely anchored in place over the window opening. To a casual observer this protection could be mistaken for conventional window bars but of course attempts to force open the tubes breaks the wiring inside them. In many instances it is not possible to adequately protect premises by relying solely upon perimeter devices and, indeed, some of the techniques employed satisfactorily in industrial or commercial premises are not suited to private homes. Few people, for example, would wish to cover every window with 'tube and wire' bars. Since in some 47% of all burglaries entry is effected through rear windows other methods of detection must be employed.

The second group of devices provide what is known as 'internal trap' protection, and aim at detecting the intruder once he is inside the premises. Such devices include pressure mats that can be concealed under carpets in front of windows or on stairs and the motion detectors previously mentioned. Three basic types are currently in use, namely: *ultrasonic* which radiate high frequency (inaudible) sound waves into a room and detect motion by sensing the change in frequency that occurs in the echo signals reflected by moving objects; *microwave* which also utilises the 'doppler frequency shift' phenomenon but radiate very high frequency radio signals instead of sound waves and, lastly, *passive infra-red detectors* (PIRs) that, as their name implies, do not radiate any energy at all. Instead they sense the infra-red energy emitted by intruders themselves and, by means of a very clever optical system, can differentiate between the infra-red signal from a moving target and the background infra-red radiation that is emitted by everything else in the room.

Although much 'cleverer' than simple door and window switches, motion detectors can and do, unfortunately, generate many false alarms, usually because they are installed in environments unsuited to this class of protection. Each of the three types react to natural phenomena in different ways so installers must choose the type of sensor best suited to the prevailing conditions. Many different criteria govern the choice of type of motion detector but, in general terms, many authorities consider that passive infra-red types are the type best suited to domestic environments.

Although all types of motion detector incorporates some form of signal processing in an attempt to minimise false alarms, for instance they may require movement to be present for several seconds or to be within 'human' speed ranges before triggering, their discriminating ability is still not much better than that of a simple switch on a door or window. A cat, for example, may 'look' the same as a cat burglar if it can pass near enough to a motion detector to provide a 'large' target. Similarly, a pressure mat or door contact cannot discriminate between intruders and the large family dog who has the run of the house.

Because of these detector shortcomings, much research and development has been devoted to the design of new generations of control panels. These so called 'smart' panels incorporate a microprocessor which is programmed to analyse the operation of the detectors and only give an alarm if certain predetermined conditions are met. Unlike existing panels that signal an alarm immediately any detector is operated, the smart panel looks for sequences or patterns from groups of detectors and if such a sequence occurs within predetermined time slots then, and only then, is an alarm generated. For example, if the panel recognised that a window detector had been activated and that this event was closely followed by inputs from, say, a pressure mat and a motion detector in the same area of the building then it would justifiably conclude that an intruder had broken in. If, however, a single detector produced sporadic inputs that were unrelated to inputs from any other detectors, then the panel would ignore them as false alarms. Such 'intelligent' operation bestows the security system as a whole with more than just a greatly improved false alarm performance. Since failure of detectors is always a possibility, whether it be due to fracture of the glass envelope in magnetic reed switches caused by clumsy installers, slamming doors etc. or a component failure in a complicated electronic detector, systems with a conventional control panel may be partially or completely out of action until the fault is corrected. The 'smart' panel, on the other hand, can accommodate some failure of detectors without depriving the property of protection. Programming of the microprocessor to recognise burglary scenarios and ignore inputs indicative of false activations may be performed by firmware, i.e. by software stored in read only memory (ROM) or, where more flexibility is required, by means of switches and ROM. The installer can then set up the device to accept any combination of sequential inputs he chooses to suit the particular property.

Comparatively little memory is required to perform this 'pattern recognition', so the panel's 'intelligence' may be put to other uses as well. A monitor programme resident in ROM can continuously check all performance parameters and, via a simple display, indicate and identify potential sources of trouble. Ease of operation is another bonus since the available capacity of the system logic easily permits the inclusion of electronic combination locks set by alphanumeric keyboards to control authorised access. This facilities relieves the home owner of the burden of extra keys for the alarm system with all their attendant security problems.

At present, all the development of such panels is devoted to 'stand-alone' units which function independently of any other electronic facilities used in the home. If, however, the security function can be integrated into 'the home computer', considerable benefits may accrue. In the very near future we shall see the introduction of home monitoring systems which will embrace all aspects of 'house keeping'. The home computer will, on a routine basis, control heating, lighting and ventilation with energy conservation (and therefore cost!) benefits; it will monitor the home environment for intrusion of the outbreak of fire and it will perform these services whilst still able to produce menus, budget the housekeeping money by directly accessing the Prestel viewdata service and so determining daily best buys in meat and vegetables; print bank statements; read your electricity and gas meters and still find time to beat any member of the family at almost any game capable of being played on the small screen.

Such developments are possible now. Indeed, computer control of building management is commonplace in the commercial field whilst integration of the security and fire functions with housekeeping operations is also growing. In the home, however, the full potential of such systems will not be realised until a second, new and revolutionary element in security and fire systems is firmly established.

Perhaps the biggest single drawback of conventional domestic security systems is that extensive wiring has to be run all over the house to link detectors with the control equipment. Such wiring is expensive to install in a home where, unlike a factory, it must be concealed from view wherever possible. A considerable degree of craftsmanship is necessary to install an unobtrusive and neat domestic security system and labour charges are rising all the time. Most alarm users rent their systems rather than buy them outright – a situation preferred by most insurance companies since a user has some redress if contracted service and maintenance visits from the installing company are irregular or non-existent. Nevertheless, the initial installation fee charged by most companies will, for many families, represent a considerable capital outlay even if the annual rental/maintenance fees are comparable with those for, say, a video cassette recorder. At present the property market appears to add little or no value to installed alarm systems so, if he moves, the householder may expect little financial return for his investment. If the purchaser of the property does not wish to take over the existing system the installing company may also lose because even if the control unit and any of the more expensive items like motion detectors are removed, perhaps as much as 70% of the systems value may have to be abandoned because it is usually impractical and uneconomic to strip out the labour intensive circuit wiring and simple contacts. When one also considers that even in domestic systems, this wiring is vulnerable to accidental damage or deliberate sabotage the concept of an alarm system without wires becomes very attractive.

Before we consider how we can achieve this desirable goal let us list a few more of the principle benefits. Firstly, installation would be minimal and therefore cheap. Secondly, because all the major elements would be easily removeable, the home owner could take the system with him when he changed his property. Equally, a security company who supplied such systems would not have to write off a high labour cost whenever a customer defaulted on the rental contract.

So how can we connect detecting devices to a control unit (preferably a smart one) without wires? Wireless is the answer! Although comparatively rare in Europe, radio alarms as they are called have been sold in North America for about eight years. With such systems each detecting device, or a small group of them, is connected to a very compact, low power radio transmitter whilst the system control unit is equipped with a special radio receiver connected to the inputs usually occupied by 'hardwire' circuit wiring. When any detector is activated by an intruder a short duration, coded radio signal is transmitted. This signal is received and decoded at the control unit which then acts upon the information in the appropriate fashion. Simple? Well, yes – providing certain very important performance parameters can be met – so why did this new technology fail to immediately replace labour intensive circuit wiring with all its drawbacks?

The principle reason why Europe lagged behind the USA and Canada and has only recently started to use radio alarms is simply one of standards. In most countries of the world, before any manufacturer can market almost any form of transmitter the equipment must be submitted for tests that are laid down by the government body responsible for radio administration. These tests ensure compliance with either national or international performance specifications and the process of testing and subsequent certification is known as Type Approval. Once Type Approved, a transmitter can be legally marketed although legal use may depend upon the issue of a licence to the user and the allocation of a specific frequency on which the equipment can be operated. Certain frequencies are reserved for particular types of transmitter or for important users. In this way each category of use is accommodated and sufficient control is exercised to ensure that vital communication links can operate without interference.

Naturally legislation governing the use of radio varies from country to country and, in general terms, geographically small countries with several common borders have to impose strict regulation to avoid interference of and by neighbouring radio services. For this reason European radio administrations frequently impose more stringent regulations than those laid down by the regulating body in the USA – the Federal Communications Commission (FCC) – whose only near neighbour is Canada. When radio alarms were first developed in the USA the FCC was able to allocate very generous blocks of frequencies (in the UHF band) and also Type Approved transmitters against a standard which, by European standards, is not very demanding. American or Canadian equipment could not therefore be used legally in the United Kingdom or in most of continental Europe, and we had to wait until new systems could be developed that complied with the regulations in force here.

The first system to gain Type Approval in this country was developed not to a UK national specification, but rather to comply with one of the recommended standards issued as by the CEPT in Geneva, much of whose work is aimed at the introduction of common standards throughout the member countries of the International Telecommunication Union. The widespread adoption of common standards will ultimately result in reciprocal acceptance of Type Approval tests and greatly facilitate export of radio equipment.

In the field of radio alarms, the product's evolution has also benefited from the advances in electronic technology that have spawned many of the other products described in earlier chapters. In some cases, microelectronic techniques have been employed to improve reliability whilst, in others, a requirement to meet more stringent specifications or to perform more complicated functions has necessitated the use of thick film hybrid devices, integrated circuits and, latterly, microprocessor chips. Without such miniaturisation the products would have been too big, too complicated and much too costly. An example of such evolution is shown in the accompanying illustrations which show, in Fig. 14.2, a radio alarm transmitter built to conform to FCC standards which uses discrete components (transistors, resistors, capacitors, etc.). By way of comparison, Fig. 14.3 shows a transmitter which conforms to European radio standards. In this instance the transmitter must be crystal controlled yet, the extremely complicated circuitry takes up no more space because major circuit elements are constructed on two thick film hybrid chips.

Although it is comparatively simple to construct short range radio links, the design of radio alarms that may be used to signal security, fire, or life safety warnings requires considerable ingenuity because reliability is of paramount importance. In this context, reliability means more than the ability to function without breakdown. It is obviously very important that any radio link used for such purposes must be able to transmit and receive coded signals and so ignore signals transmitted by any other radio apparatus. Ideally, such equipment should also be able to operate reliably even when subjected to quite high levels of radio interference.

To achieve this aim, radio alarms must employ special techniques to code the transmitter signals so that each 'message' is quite unlike those produced by other types of transmitter. This is especially important because radio alarms almost always have to operate at fre-

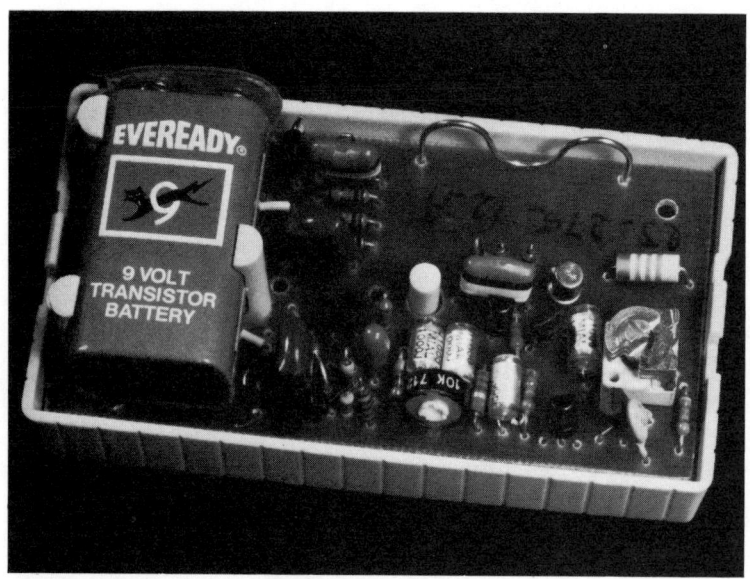

Fig. 14.2　An FCC approved radio alarm transmitter

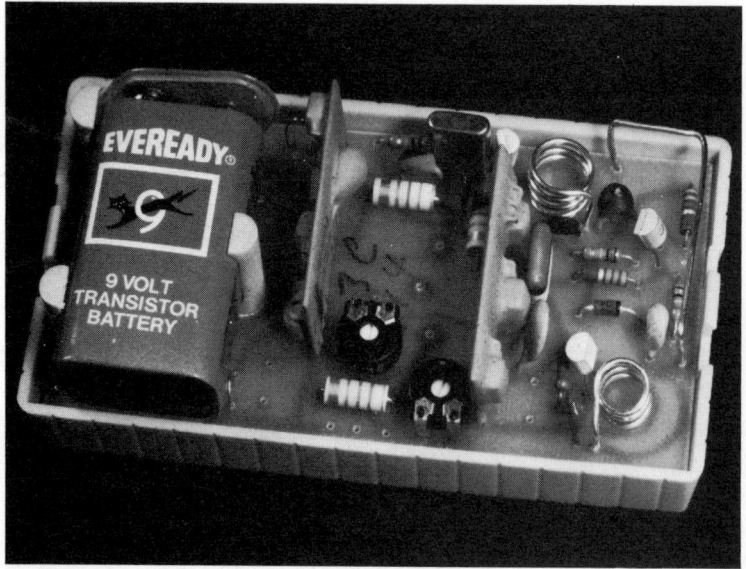

Fig. 14.3　A European radio alarm transmitter

quencies close to those allocated to other radio services, many of which employ very powerful transmitters. The techniques of encoding information on to a radio carrier wave – known as modulation – can be quite simple or extremely complex. For our purpose the more elements there are in the code the better. By way of analogy, we can liken the code elements to the notches or steps on a key. A simple key, used, say, to operate the locks on a suitcase may have only two steps whilst a high security key used to operate a lock on a safe may have seven steps and will also have a very complicated shape. If we wish to use a radio link to perform some non-critical task like opening a garage door we need many code combinations but the code can be very simple. A large number of code combinations allows very many users to have separate keys, but the code itself may not be very secure. For such applications digital techniques are often employed and each transmitter transmits a simple pulse code the combination of which can be set by the user by means of a small block of switches on the circuit board. If interference from other radio traffic prevents the receiver from decoding the message the user can always open his garage door by hand!

Fig. 14.4 shows the block diagram of a radio alarm, used for more critical applications, which is designed to tolerate high levels of interference. The transmitter is powered by a small replaceable battery and because the device incorporates a timer to limit the duration of each transmission to about one second, the battery only needs replacing every six months or so.

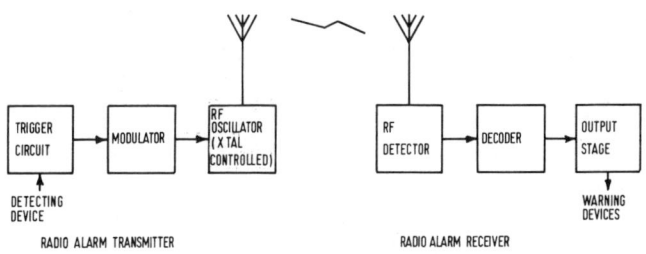

Fig. 14.4   Schematic layout of a radio alarm system

Fig. 14.5 shows a radio alarm transmitter connected to a magnetic door or window contact, whilst Fig. 14.6 shows a passive infra-red motion detector with an integral radio transmitter. This device is mains powered but is provided with an internal battery which will power the device for some 200 hours should the mains supply fail.

Fig. 14.7 shows a radio alarm control panel which receives and decodes the signals from the detecting devices/transmitters. About the size of an ordinary portable domestic radio, the control panel can sound loud audible warnings and trigger automatic telephone diallers or other remote signalling systems.

Many readers will be familiar with the self-contained fire detectors that are now marketed throughout the UK. Yet another transatlantic import, these detectors usually employ optical sensors or ionisation chambers to detect the invisible products of combustion that are generated in the insipient stages of a fire. Once triggered these detectors sound a loud built-in hooter to warn the building's occupants of their danger. Fig. 14.8 shows such an ionisation smoke detector which incorporates an integral radio transmitter. When triggered this detector not only sounds its local warning but also transmits its coded radio signal to the system control panel. As fire sensors can transmit a different code from that transmitted by security sensors, the control panel is able to identify the signal as a fire message and sounds a different audible warning. Simultaneously, a second (fire) output to the remote signalling devices is provided so that the appropriate authority (the fire brigade) is summoned.

The use of radio bestows the alarm system with another important bonus not enjoyed by hardwired systems. This is the portable transmitter which can be easily carried on the person and manually actuated whenever required. In a security system such transmitters can be used as panic or hold-up devices, Fig. 14.9. A housewife carrying such a transmitter, can trip the alarm if an intruder tries to force his way into the house and, because the control panel makes provision for reception of such 'panic' signals on a 24 hour basis, this warning will operate even if all the other burglary protection facilities are turned off. Fire warnings would also be received regardless of the status of the other protections.

Such portable transmitters have other uses in the home quite apart from those already described. Perhaps the biggest single use for such devices is the provision of a 'medical alert' system for the elderly, the handicapped and anyone living alone. Fig. 14.10 shows the component parts of such a system whilst Fig. 14.11 is a block diagram of the complete communication channel. Apart from the portable transmitter, the system comprises a receiver, a digital telephone communicator and a power pack all housed in a single cabinet. Installation of

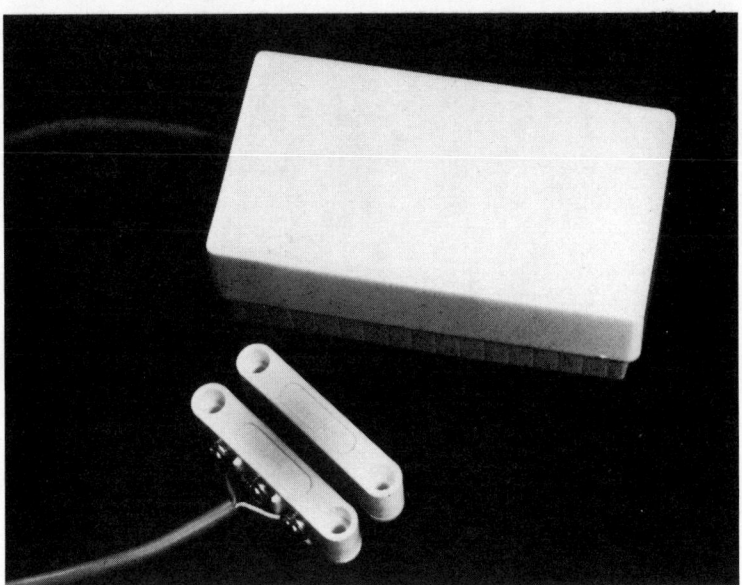

Fig. 14.5 A radio alarm transmitter connected to a magnetic contact for use on doors and windows

this latter unit is limited to a connection to the mains power supply (rechargeable stand-by batteries are built-in) and a connection to the house telephone line. The digital communicator is a modern version of the automatic telephone dialler which, when triggered, plays a pre-recorded tape message to the emergency service telephone operator after automatically dialling 999. The communicator, on the other hand, dials the number of a central station which is manned continuously by medi-alert service operators in the premises of the company operating the scheme. At the central station, the telephone lines are monitored by a special receiver which is linked to a printer and usually, a computer.

Should the housebound person require assistance he or she simply presses the transmit button on the medi-alert transmitter. The local radio receiver then initiates the digital communicator transmission sequence which operates as follows: After sensing that the telephone line is clear (by detecting the dialling tone), the communicator dials the central station number. The remote receiver acknowledges the incoming call by transmitting a signal known as a handshake. This signal tells the communicator to go ahead with its message which it

Fig. 14.6   A passive infra-red detector with integral radio
transmitter

transmits in digital form at high speed. The message code is repeated
several times and the receiver compares each message text to check
for errors caused by noise on the line, bad connections, etc. If three or
four error free messages are received the central station unit transmits
a 'kiss-off' code to the communicator which tells it that the message
is received and understood and that it can now hang up. The message
code is then printed out, together with the date and time, so that the
operators can check the origin of the message in their log and take the
appropriate action. If linked to a microcomputer, reception of a message by the central station equipment causes a search through the
memory files for details of that users status. This information, together
with the action necessary is then displayed on the VDU of the
computer.

Fig. 14.7 Control panel designed to receive signals from radio alarm transmitters

Such a text might read:

Attention: Attention: Attention
Medical Alert. Time of origin 02.37 a.m. 21-12-79
Subscriber: Mrs DM Jones, File 913
Address: 23 Accica Drive, Chelsea SW3
Telephone number: 01 581 00317
Status: Severe Arthritic – Limited Mobility
         Age 79. Partially Deaf
         Lives alone. No next of kin.

ACTION: Call 'Good Neighbours'
Mrs Ada Goodchild, 21 Accica Drive, telephone 01 581 99134

243

Fig. 14.8   Ionisation smoke detector with integral radio alarm transmitter

Fig. 14.9   Radio alarm 'panic' transmitter

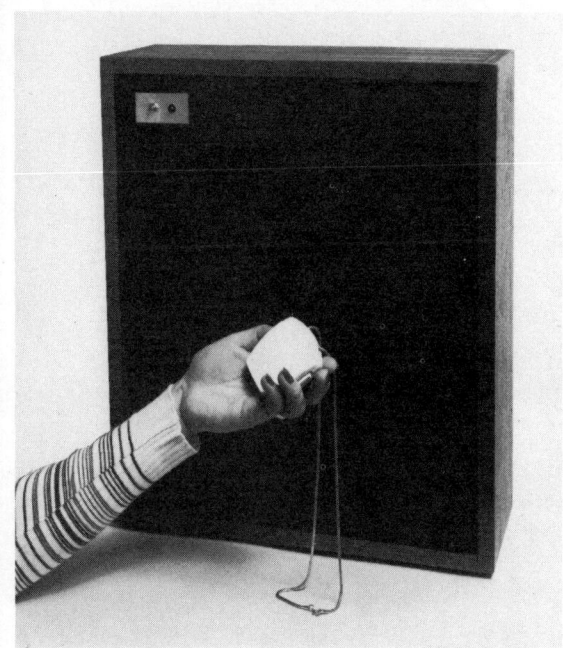

**Fig. 14.10   Miniature Medialert transmitter with the
Medialert receiver/digital communicator**

Alternative: Mr Michel Wright, 25 Accica Drive, telephone 01 581
02753
Doctor: MD Parkes, telephone 01 581 53254
Nearest Hospital: St Peters, telephone 01 584 21970

Remarks: Be advised, last call 3-10-78. Hospital admission – hypo-
thermia.
ENTER: action taken and results. File 913.

END

In concept, such systems as described above are not new, but most
alert systems rely upon the installation of hardwired call buttons at
strategic locations within the house or flat. Apart from the high cost of
installing such call buttons there is always the chance that, because of
a fall, an elderly person would not be able to reach one of them. The
alert transmitter is easily carried or may be worn around the neck like
a pendant. The costs of subscribing to such a service – which in any

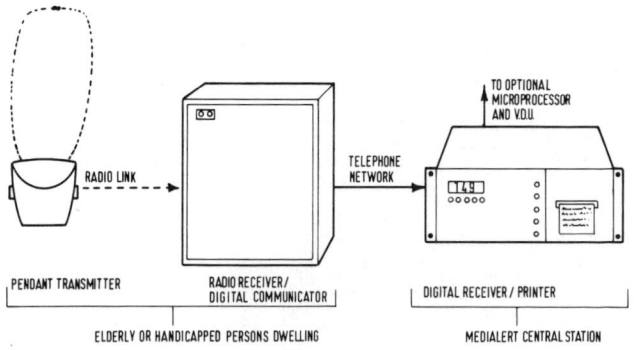

Fig. 14.11 The components of a medical alert system

event may be borne by local authorities or charitable bodies with special responsibilities for the elderly, the handicapped and the sick, are comparable with the rental charged for colour televisions, video cassette recorders and so on.

Within a book such as this, it is not possible to describe more than a few of the uses that such equipment is now finding in the home, but mention must be made of the new generations of radio telemetry and telecommand systems that will shortly be making their debut. In these devices, custom designed integrated circuits that draw heavily upon experience gained in microprocessor applications will be used to greatly extend the versatility of radio alarm systems. Such chips will allow each transmitter to individually address a receiver with thousands of available codes whilst still retaining a very high noise immunity and freedom from RFI false alarms. Each transmitter can be programmed to select one of several different alarm messages, according to the input it has received, whilst regular 'test' messages inform the monitoring receiver that the communication link is open. The batteries in each transmitter are constantly checked and a 'battery low' code is sent as soon as an out-of-limit condition is approached (Fig. 14.12).

The widespread use of these low power, short range radio links, engendered by prices which, already low, will fall still further as production increases, should add considerable impetus to the electronic revolution that is just beginning in our homes. What are currently regarded as technological miracles will undoubtedly be commonplace tomorrow. Such devices can, if applied sensibly, improve the quality of life of people in all walks of life and of all ages. A broadening of leisure activities, security against crime, fire and accident, improved communications – when the silicon chips are down, almost anything is possible and we've only just begun!

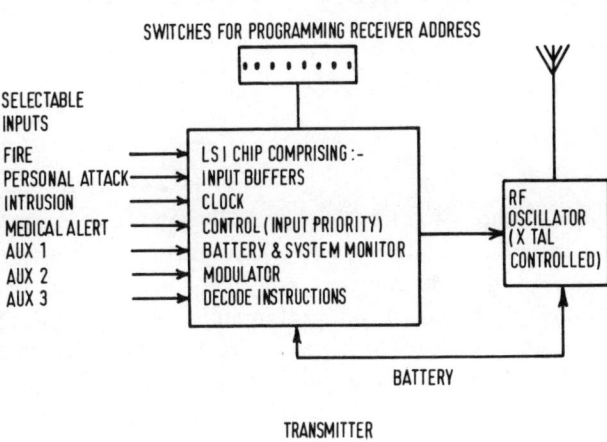

SWITCHES FOR PROGRAMMING RECEIVER ADDRESS

SELECTABLE INPUTS

FIRE
PERSONAL ATTACK
INTRUSION
MEDICAL ALERT
AUX 1
AUX 2
AUX 3

LSI CHIP COMPRISING :-
INPUT BUFFERS
CLOCK
CONTROL (INPUT PRIORITY)
BATTERY & SYSTEM MONITOR
MODULATOR
DECODE INSTRUCTIONS

RF OSCILLATOR (X TAL CONTROLLED)

BATTERY

TRANSMITTER

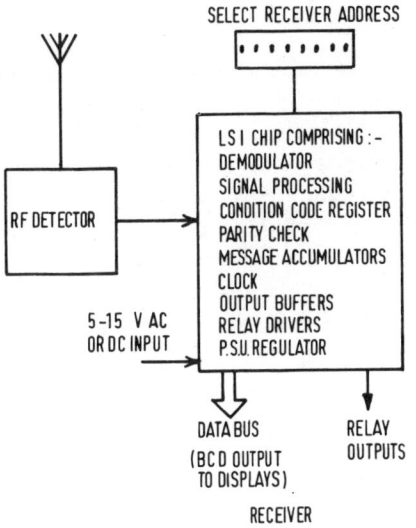

SELECT RECEIVER ADDRESS

RF DETECTOR

LSI CHIP COMPRISING :-
DEMODULATOR
SIGNAL PROCESSING
CONDITION CODE REGISTER
PARITY CHECK
MESSAGE ACCUMULATORS
CLOCK
OUTPUT BUFFERS
RELAY DRIVERS
P.S.U. REGULATOR

5-15 V AC OR DC INPUT

DATA BUS
(BCD OUTPUT TO DISPLAYS)

RELAY OUTPUTS

RECEIVER

Fig. 14.12   Schematic of 'second generation' microprocessor based radio alarm system

247

15   Amateur and Citizens Band Radio

Richard Elen (G8RJX)

Although Citizen's Band (CB) and Amateur Radio have a great number of things in common, there are also some fundamental differences, primarily in terms of the legal position (as regards licensing) and the uses to which these two types of radio communication are put. So, first of all, we'll take a look at Amateur, or 'Ham' radio and then examine CB. But remember that the conditions under which amateurs and CB'ers operate vary in different parts of the world – if you're interested in either form of communication, check with the authorities if you live outside the UK before you run the risk of infringing the licensing regulations.

Amateur radio

Most of us probably imagine the 'ham' radio enthusiast as locked away in a tiny room surrounded by glowing valves and electronic equipment, speaking across the world to crackly, distorted voices emerging from battered headphones. But there's far more to amateur radio than this Tony Hancock concept. Modern ham radio is a sophisticated art, offering high-quality long-distance communication to many thousands of individuals who are interested in radio, not from a commercial or business point of view, but purely because they enjoy either the technical aspects of the hobby or the thought that they can meet, over the air, a new friend with every contact, whether they're talking to the other side of the world or just down the street.

This indeed is one of the thrills of amateur radio, and many who cannot get out and about for one reason or another are a part of the band of quite ordinary people who meet others they'd never know existed, through the medium of radio. Many people use their equipment from home in their spare time, and others have a mobile 'rig' in the car – or even pushbike – and keep each other company on the journeys to and from work, or on business trips, as well as exchanging traffic information and the like.

But despite the apparent sophistication, it's hardly difficult to get into amateur radio, and hundreds get their licences in Britain every year. It's a simple matter to obtain a licence – in the UK, all you have to do is to pass a simple examination and, if you wish, a Morse test, and fill in a form.

The Radio Amateur's Examination may sound a daunting concept, involving as it does questions on physics, electronics and radio, and requiring a knowledge of the licensing conditions imposed by the UK Home Office. But it really isn't that bad. Anyone who's done secondary school Physics needs little more to pass, and many local radio societies and colleges offer evening classes for intending amateurs which last only a few weeks. The Radio Society of Great Britain publishes a book, the *Radio Amateur's Examination Manual*, and they suggest that if you were to learn the slim volume off by heart you'd pass the exam! A Home Office leaflet, *How To Become A Radio Amateur*, explains the licensing side of things, and the combination of these two booklets plus a short evening course will make a pass a virtual certainty for anyone over the age of 16. And it's not purely a male hobby either!

The exam is held in a number of centres throughout the country twice a year, in May and December, and consists of a multiple choice question paper. After obtaining a pass, it's a simple matter to send up your form and obtain an Amateur Licence. A simple exam pass will enable you to obtain what's known as a 'B' Licence, which means you can operate on the VHF amateur bands (144MHz, '2 metres', and above), but if you learn the Morse Code in addition you can get the full 'A' licence and operate on the short wave bands as well. Morse code classes are again held by many local societies.

With the licence comes a 'call-sign', an individual group of letters and numbers consisting of a letter or two indicating the country (G, for example indicates England; GM, Scotland, and so on), followed by a number indicating the class of licence, and then a group of usually three letters issued in alphabetic sequence. This callsign serves to identify the user, and is modified by an extra letter to indicate whether the operator is at home (just the simple callsign is used), mobile (/M), or portable for example using a walkie-talkie (/P), or in a temporary location (/A). Thus the callsign G8AAA/M would indicate that the user was a G8 ('B' Licence, VHF only), and in a car.

Whilst the short wave bands offer far greater scope in terms of the possibility of working long distances, the basic G8 call isn't such a limiting thing. There are many amateur bands available, even above the 144MHz lower limit of the 'B' Licence. Most popular is the '2-metre' band, 144-146MHz, and whilst this (and the higher frequency bands) theoretically offer only 'line-of-sight' transmission and reception, under certain 'lift' conditions contacts can be made over hundreds of miles with only a few watts output. Then there are the 'repeaters', which receive on one frequency and transmit on another, 600kHz above the receive frequency. There are many repeaters spread over the country, mounted on local high spots. These increase the effective range of mobile stations and those in bad locations, as well as enabling amateurs, say in hospital with only a low-powered rig, to keep in contact with the outside world. The same situation exists on the '70-centimetre' band, around 430MHz, where there is far less activity.

But the short-wave, or 'HF' bands as they're called, open up a whole new field, literally a world full of potential contacts. The amateur fraternity throughout the world is no less than a large group of potential friends. It is not an elite – it is a fraternity that is easily joined, by fulfilling the requisite conditions. Licensing conditions are quite strict, of course – they have to be, to avoid interference and misuse. You can only transmit on certain specified bands, and with certain maximum power levels, to avoid interference with other services; thus there is a great deal of experimentation with better aerials and transmitting systems to make best use of the bands whilst remaining within the law. In fact it was amateur experimentation that opened up the short wave and VHF bands for the first time, when the authorities thought much of the spectrum was useless. And, of course, it's quite a challenge to try and contact someone over a long distance with very low power. All for the price of a few hours learning, sitting an exam and paying the Home Office a few pounds.

Citizen's Band

Citizen's Band – or CB – is a different kettle of fish. It began in the USA; it was first licenced in 1949, but did not come into the public eye to a great extent until the trucker's strike a few years ago, when truckers used CB to co-ordinate industrial action; today over one in four US road vehicles have CB sets, and some manufacturers are even considering fitting them as standard units, just as you might expect a car radio in Britain. Since 1949, CB has spread to many countries, including Canada, Germany, Australia and Italy; even behind the Iron Curtain, there is a CB in Poland and Czechoslovakia. In most of these countries it is regarded, sensibly enough, as merely a part of the right of freedom of speech. People can be trusted to run free newspapers and so on – why shouldn't they be able to be trusted with the modern technology of radio communication? So the argument goes – unfortunately it is one which is totally denied by the British Home Office. Britain is quite unusual in its failure to introduce a CB. Arguments against it include suggestions that it encourages crime (which it doesn't – how can you organise a crime when everyone else can hear you?), that there aren't enough spaces in the British radio spectrum for it (although the USA, which has far more radio use than the UK, and in the same band-space, has managed not only to find room, but also to expand the number of CB channels from 23 to 40, and allocate an entirely new, UHF CB), and several others. On the other hand, American users, and British pressure groups, point out that it can save the lives of, for example drivers stuck in snowdrifts who can put a call out for assistance on the Emergency Channel. CB is even endorsed by the USA police, who say that far from encouraging drivers to break the law (particularly the speed limits), it appears that these days, all that is needed is for a police car to be seen and people will be on the radio telling everyone else for miles around to slow down! Yet it remains illegal in the UK – thus encouraging illict use by many people (over 1,000 in London at the end of 1978) who have managed to get hold of, or bring back, an American CB rig, despite potential heavy penalties. A 40-channel CB rig in the US can cost as little as £20; in comparison, a 2-metre amateur unit can cost you as much as £300.

American CB activity is primarily on the 27MHz short-wave band – in fact this is a bad choice for Britain, as it can cause interference to Band I TV. CB groups in the UK have opted for VHF/FM, much along the lines of 2-metre amateur gear; a much better solution. But the remarkable thing about CB is its accessibility. In the US, all you do is buy a rig, put it in the car, or at home, fit a good antenna, send off a few dollars to the FCC and you're away. No exams are needed, and the restrictions are few. Everyone, almost, can get in on the act – appropriately enough, in the context of free speech. Hardly surprising that there's so much interest in Citizen's Band. Not only is CB available simply for a chat with someone you'd never otherwise meet – it also provides a highly cost effective method of fitting small companies with two-way radios, who perhaps cannot afford the far more expensive private mobile radio equipment, and, maybe, don't need that kind of service. Then there are the other advantages of instant traffic reports, emergency calling, and the like. Thus it is that several organisations in the UK, including the National Electronics Council, chaired by Lord Mountbatten and representing radio manufacturers, educational establishments and the like, have been trying to persuade the British Government to introduce CB – so far, regrettably, with no success. You may think it strange that a democratic country is without this important adjunct to freedom of expression; there are, after all, few, if any sound arguments against it.

The access to CB also introduces people to modern technology and would be likely thereby to educate people in its use; in fact in the US many CB'ers have taken the next step – they've taken the amateur examination and moved up into the realms of world communication from their own homes; world communication which brings nations and peoples together and knows no political barriers. A small, but important step towards world peace; a step that, through amateur radio, and indeed CB, is, or could be, open to all.

16   Hi-Fi Developments

John Atkinson

If you think about it, it was sound reproduction that triggered off the whole consumer electronics industry. People had always been fascinated by the idea of being able to preserve a man's voice or music in a form that could be recalled at will. It was just 100 years ago, in 1877, that a Frenchman, Charles Cros, and an American, Thomas Alva Edison, almost simultaneously hit upon the idea of preserving sound waves as a squiggly line scratched in a softish substance which would then make anything pulled along the scratch vibrate and reform the sound. Cros never developed his idea but Edison, who was by all accounts a one-man NRDC, produced a machine which scratched a spiral groove on a tinfoil-coated cylinder, preserving for posterity Edison shouting 'Mary had a little lamb' at the top of his voice as a series of hills and dales in the bottom of the groove.

Very rapidly an industry grew up to exploit the Edison phonograph which used wax cylinders, recorded and played back purely acoustically. Cylinders were eventually replaced by flat discs but sales were plummeting by the early 1920s and it looked as though the public had tired of the novelty, most probably because of the limitations on both quality and quantity of sound imposed by the primitive level of technology applied. Many ingenious ideas had been developed, including an acoustic amplifier operating on compressed air, but it was the invention of the triode valve, the first electronic amplifying device, by Lee de Forest in 1907, and the loudspeaker by Rice and Kellogg, which revitalised both sound recording and reproduction.

The importance of the application of electronics to recording can be judged by the fact that in 1925 HMV had recorded Elgar's 2nd symphony acoustically. It was a very expensive project to record a symphony, even in those days, and yet within a matter of 18 months, they went to the trouble to do it all over again, inspired by the improvement in quality offered by the use of a microphone and amplifier to cut the record rather than what had been in use previously, a very large horn attached to a diaphragm and stylus, around which the musicians had had to crowd.

The triode had come from the world of radio and the late twenties and thirties saw an explosion in the interest in recorded sound, fuelled by radio, records and the cinema, all of which utilised and forced development in a common technology. The next decade and a half saw a great deal of tinkering with the processes giving small improvements, but, essentially, people listened to; mono 78rpm records, with the signal cut *laterally* i.e. snaking from side to side, on either an acoustic or an electric gramophone; AM (amplitude modulated) mono radio; and mono film soundtracks, recorded as a band of varying light and shade on a strip running alongside the frames on the film.

None of these processes produced what could be defined as high fidelity. The human ear can hear sounds in a frequency range of approximately 20-20,000Hz, depending on age, and in a range of loudnesses in a ratio of 1,000,000 to 1. True high fidelity would mean reproducing sound in such a way that a listener with his (or her) eyes shut would be fooled into believing that he (or she) was witnessing the original event. We still have a long way to go on that score but a small number of major developments took place in the years following World War II which brought high fidelity within reach. Most fundamentally important was stereo. A mono recording can be regarded as analogous to a black and white photograph; the form is there and most of the relationships between objects are correct but it is very much a map of a territory rather than the territory itself. Stereo, by using two channels, adds the colours and attempts to give an idea of depth. It still isn't the territory itself but rather a good model instead of a map.

As far as the disc is concerned, all the basic engineering required and the philosophy required had been worked out in the early thirties by a brilliant engineer at the EMI laboratories in Middlesex, Alan Dower Blumlein. Some of his stereo recordings are still around and sound very impressive, but the record industry just wasn't ready for such innovations, and Blumlein's ideas went into cold storage. Blumlein, incidentally, was killed in a wartime plane crash: had he lived, I'm sure that he would now be widely regarded as one of the leading scientific figures of the 20th century. The problem involved in recording a stereo signal on a disc record and its solution will be covered later in this article but, combined with the invention of the microgroove longplayer by Dr Peter Goldmark of CBS in America in the early fifties, it produced the record as we know it – a stylus being waggled in a groove in a 12in PVC disc revolving at 33⅓rpm. When CBS introduced the LP, their great rivals, RCA, brought a rival system based on a record 7in in diameter revolving at 45rpm (a much more optimum speed, fidelity-wise, by the way) but the race went to CBS, with all the record companies adopting it (even though Decca, for instance, had almost got their own system ready) and the RCA disc now only lingers on as the 'pop' 45 single.

Radio took its big leap forward with the invention of the FM (frequency modulation) system by Armstrong in America and its adaption to stereo broadcasting; again an idea developed years before its commercial exploitation. Film soundtracks carried on using the relatively primitive optical system but the unforeseen advance was the invention of the magnetic tape recorder, most of the technical development work being carried out in Germany before the war but the commercial development taking place later on in the USA with Ampex, financed by singer Bing Crosby, taking the lead.

Fig. 16.1   The good old days

So now we have all the ingredients of hi-fi systems as they exist to-day: the disc with the basic idea over 100 years old and its application to stereo 45 years old; the amplifier which is basically still, underneath all the tinkering, de Forest's triode; the radio, either AM – 60 years old – or FM and FM stereo – 30 years old; a tape machine – basic principles around 40 years old; and the loudspeaker, the most popular kind – the moving-coil unit – dating from the 1920s. I will now examine each in more detail and take a look at the future (assuming, that is, that Western Society will continue into the next century).

The loudspeaker

Everyone is familiar with loudspeakers – from the smallest 'tranny' to the monster public address systems used by groups these days; all are based on a transducer changing an electrical signal into an acoustic one and probably 99·9% of loudspeakers use a coil, carrying the electrical signal, moving in the gap between the poles of a circular permanent magnet. One end of the coil is attached to a cone, the edge of which ends in a flexible surround, and this drives the air. This was the principle invented by Rice and Kellogg all those years ago and you may be understandably wondering why this one basic idea has spawned such a diversity of shapes and sizes.

Well, without beating about the bush, the moving coil loudspeaker is a very imperfect device – unfortunately, with two or so exceptions, which have their own peculiar disadvantages, nothing has come along to replace it. Many of the problems derive from the fact that sound from the rear of the cone is moving in the opposite sense to that from the front. If the two are not prevented physically from meeting each other, they will interfere destructively resulting in no sound (actually, as the interference is frequency dependent, the bass performance is most affected). So, we put the speaker in the middle of an extremely large flat baffle and that solves that – or would solve it if we weren't concerned about putting our speakers in rooms rather than playing fields (Fig. 16.2). So we apply a little ingenuity and fold the baffle into a sealed box. Result: a manageable speaker unit and the sound from the front and back of the cone is still prevented from interfering. This is the so-called 'infinite baffle' enclosure and is often referred to as an 'acoustic suspension' type because the springiness of the enclosed air mass controls the cone motion. Unfortunately, a speaker of this type has a bass extension directly related to the box size – it falls off below the resonant frequency of the cone mass and suspension compliance

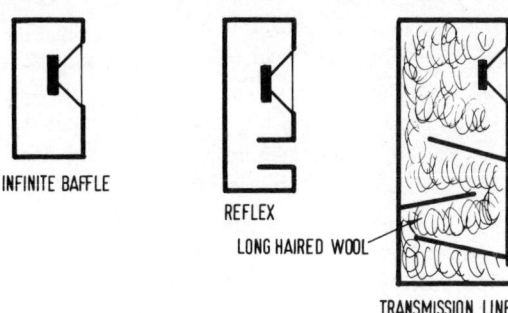

INFINITE BAFFLE

REFLEX

LONG HAIRED WOOL

TRANSMISSION LINE

Fig. 16.2   Basic enclosure types

– and efficiency also suffers. If we load the back of the cone with a long tapering tube fitted with an absorbent material (long-haired wool has been found to be best) we have a 'transmission line' (from the electrical analogy) and solve the bass extension problem. However, efficiency is very low. If we arrange for the rear radiation to reinforce that from the front via a tube or port in the baffle, we increase the efficiency and lower the bass extension somewhat but lose some control of cone motion. A variation on this reflex design uses an ABR – auxiliary bass radiator – where the air mass in the port is replaced by a passive cone/ suspension system, but the basic principle still remains the same.

All three types of enclosure have their own advantages and disadvantages and, if well engineered, can provide satisfactory performance. Slightly different is the idea of loading the front of the cone with an exponentially flared horn. This has the effect of a transformer in matching cone motion to air motion and is very much more efficient, around 40% compared to 1-2% for an acoustic suspension type. Hence their widespread use in the days when high power amplifiers didn't exist and in PA systems where you need to squeeze every drop of audience-stirring volume out of the equipment. Because of the high efficiency, the cone hardly needs to move to produce reasonable volumes so transient, distortion and pulse performance can be very good. However, horns have their own weaknesses; in particular, the size of the mouth is directly related to the cut-off frequency and a horn with the same bass extension as a modest IB enclosure is very much larger. Probably the most commercially successful horn design is the tweeter used in the Tannoy dual concentric unit.

There isn't space to go into such exotic but ultimately unsuccessful

257

devices such as those which modulate an RF arc, but the two types mentioned as exceptions earlier deserve a look and are increasingly getting attention paid to them by the Japanese. Some 22 years ago, the Acoustical Company (Quad) in the UK introduced an electrostatic speaker (Fig. 16.3) in which a plastic sheet moved uniformly between two curved perforated charged plates. Because the whole area is driven uniformly and the moving mass is very small, distortion is very low and transient reproduction is very good. Quads are bulky and can cause interface problems with rooms, but they remain one of the world's reference loudspeakers even today. Mike Oldfield, a musician known for his insistence on high quality sound reproduction, used 64 for his latest tour. Ribbon loudspeakers use a very light aluminium ribbon suspended between the poles of a strong magnet and share many of the advantages of electrostatics. They do seem to be very expensive to manufacture, though – a new Technics tweeter

Fig. 16.3   The Quad Electrostatic speaker

costs the equivalent of £160 and recently several designs have appeared, from Wharfedale, in particular, which have attempted to combine the advantages of ribbons with the ease of manufacture of moving coils by using a flat diaphragm with a flat coil printed on it. Perhaps the most expensive commercial speaker system in the world, the £11,000 (with amps) USA Mark Levinson system uses Quad electrostatics for mid-range and upper bass, Decca ribbons for treble and 24in moving coil units for lower bass. The ultimate ribbon speaker was supposedly put together at MIT when students suspended an aluminised mylar sheet between the poles of their cyclotron magnet – they apparently have not been satisfied with normal speakers since!

Design aims

All the above are just ingredients – how are they combined satisfactorily? This is the area where a state of flux exists and sometimes bitter controversy can rage about two different solutions to the same problem. Loudspeaker aberrations can be roughly generalised into two areas – those affecting the amplitude domain, such as departures from flat frequency response; and those in the time domain. The latter include the distortion given to pulse behaviour by the acoustic centres of drive units handling the same signal being differing distances from the listener. This was the problem which the time aligned, the mis-called 'linear phase' loudspeakers (Fig. 16.4) introduced originally by Technics and B&W, were intended to correct. Although the argu-

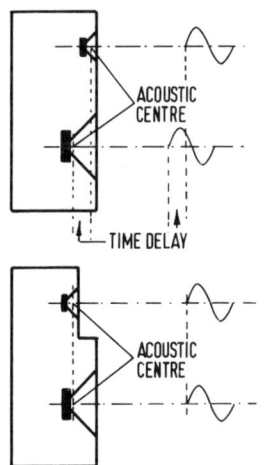

Fig. 16.4 Basic 'linear phase'

ments still rage, there does appear to be a tendency for such designs to have better stereo imagery, which would be accounted for by better pulse behaviour, than otherwise if all things were equal. However, the improvement offered appears to be marginal as more conventional speakers can be just as good if not better in those areas where 'linear phase' speakers are supposed to score.

A perhaps more promising development, also in the time domain, is the effect of energy storage in the enclosure and the baffle in particular. Speakers such as the small Rogers LS3/5A BBC monitor have excellent imagery, partly due no doubt to the tiny stiff baffle. More important however, is the image smearing due to the diffraction of the sound waves at the edges, an effect only taking place at highish frequencies due to the small wavelengths involved. The KEF design team

Fig. 16.5   One of the best regarded UK designs –
the Spendor BC1 which perhaps typifies the
'BBC' sound

Fig. 16.6   KEF R105 – computers were used extensively in its design

in the UK have been using computer simulation of this and other loudspeaker behaviour patterns, and their impressive Model R105 (Fig. 16.6) has the mid and HF drivers set in very small baffles with rounded corners to optimise such diffraction. Energy storage in the enclosure at low frequencies can be minimised by making it either as rigid as possible with cross bracing etc., or loading the cabinet walls with mass in the form of lead impregnated bitumen pads. It is interesting to note that the Quad electrostatic speaker, as well as having a very linear (i.e. low distortion) drive mechanism, has superb imaging, is time aligned and, having no baffle, doesn't suffer from the above problems.

Of course, I have been glibly talking about HF, mid and bass units without going into why several different drive units are needed to cover the range 20Hz to 20kHz. Unfortunately, the laws of physics dictate such different modes of behaviour for units to operate at extremes that it is only really practical to design a unit to cover a few octaves only. It can be kept reasonably aberration-free in this range – above and below it, a crossover system feeds the signal to other limited band-width units covering different ranges. Nearly all loudspeakers have used passive L-C-R networks operating at loudspeaker drive signal level to direct the correct fractional signal to the correct unit, but this does introduce a host of electrical and response problems and sur-prisingly only recently have designers looked at the possibility of doing the frequency splitting required at line level i.e. before the signal reaches the power amplifier. It is much easier (and cheaper) to tailor a crossover to the required drive units at that level; the only disadvan-tage is that each drive unit then requires a separate amplifier. How-ever, with falling electronics components costs, it will only be a matter of time before such 'active' loudspeakers become very much more widespread – systems such as the Linn Isobarik, Meridian M1, Rogers Reference Monitor and Griffin 85 in the UK are at the moment aimed at the top end of the market but in hi-fi, where the top end leads, the mass market usually follows and in the most problematical crossover range, that of the extreme bass, there is already a medium priced active system available from JR.

With an active loudspeaker, there is also the possibility of attempt-ing to control, electrically, any deviations from ideal cone behaviour. This, in a nutshell, is the essence of the Philips Motional Feedback system (Fig. 16.7), but it seems that there is a large inertia involved in

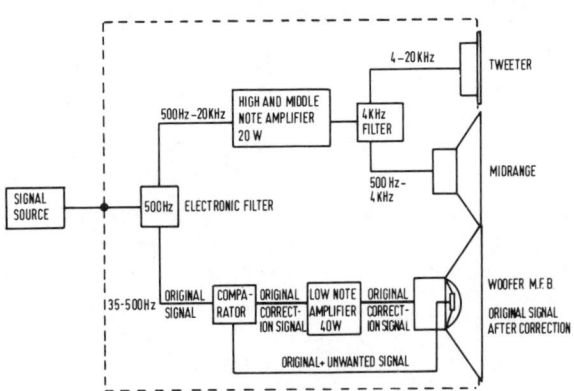

Fig. 16.7   Philips Motional Feedback loudspeaker

trying to convince the public of the efficacy of anything so radically different. The idea has since been developed one stage further with a new sub-woofer from the Swedish firm 3D Gruppen, who market their products under the name Audio Pro. Here the amplifier behaviour is altered so as to change, electrically, such factors as cone mass and suspension compliance and damping to get optimum performance at every frequency down to a low 20Hz.

The world scene

I have kept very much to the work being carried out in the UK and you may well be wondering what has been going on in the USA and Japan in particular. I do feel, though, that UK designs have had one great advantage in that, thanks to the BBC, they have always had a single design philosophy in view, that of producing a loudspeaker with as little effect on the program as possible. Elsewhere – remember please that the restricted space available unfortunately leads to rather sweeping generalisations – while drive-unit research has gone on at a frantic pace (particularly in Japan), the overall achievement of reality has not been a particularly important design aim.

Perhaps the most influential ideas in loudspeaker design have emanated from the west coast of the USA, where the impetus given by the needs of the motion picture industry enabled a great deal of the fundamental work to be done very early. Firms like JBL produced speakers based on a philosophy of high efficiency and power handling (Fig. 16.8). They produce an impressive sound with good transient performance and the world in general has followed in preferring these speakers but, in general, one important aspect of reality, flat frequency response, has been somewhat sacrificed to get nearer other aspects, dynamic range and transient reproduction.

The East Coast firm of Acoustic Research pioneered the acoustic suspension idea and made speakers sounding around halfway between the UK and West Coast sounds, but in the late sixties they carried out a good deal of work on the speaker/room interface. Their recent AR9 (Fig. 16.9) uses the room effect to obtain adequate bass response. Involved in the research was engineer Roy Allison and a new firm, Allison Acoustics, produces a range of speakers which are intended to interface constructively with the room rather than haphazardly as do other speakers.

Fig. 16.8   JBL L212 which uses an active bass unit

The future of loudspeakers

Unless anything startling happens, we can expect to see a series of
further refinements along the lines outlined above. The moving coil
speaker will continue to reign, although a new Quad electrostatic will
appear soon and a small number of wide range ribbons may be
introduced, despite the massive magnets required.

Amplifiers

The history of amplifier design is basically one of refinement and
gradual improvement. Amplifiers can, for simplicity's sake, be regarded as three separate components – the preamplifier which includes the tone and volume controls, the amplifier to boost the disc
signal to a level comparable to that from tuner or tape, and all the

switching arrangements; the power amplifier, which takes the 500mV or so from the preamp and amplifies it to a few tens of volts to drive the loudspeakers; and the power supply.

The biggest single event to happen in the evolution was the introduction of the transistor and although early solid state designs suffered from poor devices and shortsighted design philosophy, it has been accepted nearly everywhere that the transistor, of one sort or another, is the device of the future. The last three years or so, though, have seen a resurgence of interest in the thermionic valve and a number of expensive new valve power amps have appeared from TVA, Videotone, and Radford in England, Futtermans in the States, and Lux in Japan. Advocates have put forward persuasive arguments for the improvement in sound quality said to be conferred by the use of valves, argu-

Fig. 16.9 First in a new design concept from Acoustic Research, the AR9 uses two woofers set close to room boundaries to help maintain bass performance

ments invoking the subjective effect of distortion component spectra, the improvement in headroom and the effect of 'soft' clipping as against the transistor's 'hard' clipping. However, to work with low impedance loudspeakers, a valve amplifier needs a very high quality output transformer in order to preserve good performance at frequency extremes and these are of necessity very expensive. Electrostatic loudspeakers, however, operate on high voltages and one design by Acoustat in America couples electrostatic panels directly to the anodes of the output valves to form an impressive active speaker. However, the debate was summed up by a US magazine in that electrons do not remember the devices they have been through, only the quality of the engineering.

Recently, new solid state devices have appeared, using variations on the field effect transistor (FET) adapted for power amp use. The first to appear was the VFET (vertical FET) from Sony and Yamaha, which has a valve-like transfer characteristic but more promising is the power MOSFET developed by Hitachi. This is a far more linear device than the normal bipolar transistor and makes it that much easier to achieve good performance; in fact the specification of the first amplifier to appear using these devices is stunning. A pointer for the future is that several semiconductor firms are working on their version of a power MOSFET and several UK manufacturers have developed models using them.

Devices are only part of the story, though. They can be used in a number of circuit configurations, all given names such as push-pull, Class-A, Class-B, Class-A/B (not unexpectedly a combination of A and B), Class-C, Class-D, Class-E (or G), Class-A +, Class-H, Current Dumping (or Feed-forward) etc. etc. Historically, Class-A came first; the devices pass current all the time whether there is signal or not and it is therefore very inefficient. Class-B uses the devices so that they only pass current when signal passes – unfortunately transistors aren't linear at their turn-on points and the result is a particularly nasty distortion called crossover distortion. Nearly all transistor amplifiers at the moment use Class-A/B where the devices operate in Class-B except that a small continuous Class-A type bias current prevents them from being turned off thus keeping them on the linear part of their transfer characteristic.

Some interesting circuit ideas have been utilised in commercial designs to appear in the last three years, some completely new and some old ideas which have only really become recently practicable.

One of the wholly new circuits, 'current dumping', is due to Michael Albinson of Quad and is used in the Quad 405 (Fig. 16.10). This uses a Class-A amplifier to provide the required voltage swing but an ordinary Class-B amplifier to provide the necessary current. The only current required from the high quality Class-A amplifier is to modify the Class-B amplified signal current so that it conforms to the input signal – an elegant idea indeed. First of the old/new ideas is the PWM (pulse-width-modulation) or Class-D amplifier (Fig. 16.11) introduced in 1978 by Sony. This chops the input signal using a very high frequency squarewave and codes it as a series of varying length square pulses. Output current is then proportional to the mark/space ratio. The result is very high efficiency, theoretically no heat need be dissipated in the amplifier at all, but you do need very fast transistors to provide the high frequency switching and heavy shielding is required to prevent interference with medium wave radios.

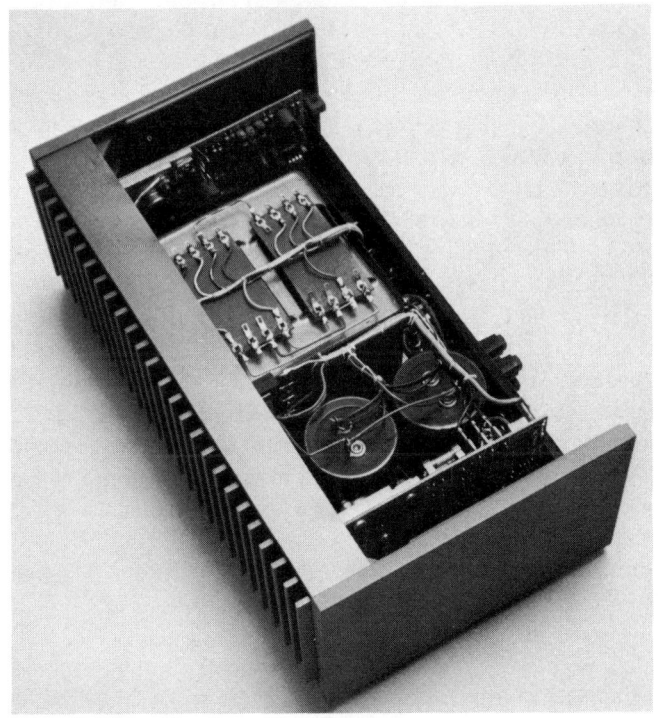

Fig. 16.10 Quad's 405 current dumping amplifier. Most of the interior is taken up by the power supply

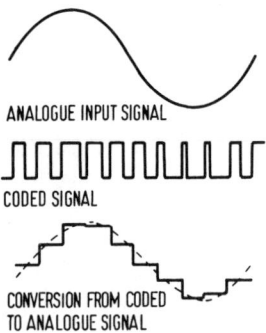

ANALOGUE INPUT SIGNAL

CODED SIGNAL

CONVERSION FROM CODED
TO ANALOGUE SIGNAL

Fig. 16.11   Pulse width modulation amplification principle

Another idea is also aimed at increasing efficiency. Class-G (or E or H or 'Dynaharmony') is used in amps from Hitachi in Japan and Soundcraftsmen in the USA. This utilises two high voltage rails rather than one as is usual. With normal level signal, the lower rail feeds the output devices but if a high amplitude transient comes along, switching diodes bring the higher voltage rail into operation, avoiding clipping of the transient. The idea is not new but the difficulty existed in switching fast enough between the voltage rails. The Japanese have perhaps been most ingenious of all amplifier designers – it seems that every major manufacturer has at least one product different from anyone else's. Hideki Ohara, chief designer at Kenwood (marketed under the name Trio in the UK) has been pursuing a philosophy involving very high slew-rate ($170V/\mu s$) power amps (the slew-rate obtained with a new device, the Emitter Ballast Transistor which is really 300 small transistors in parallel on one substrate in one package) with a response almost down to DC, connected to the speakers with special leads no more than 3ft long. The general idea is for the amplifier to have much better control over speaker behaviour and to have a power bandwidth as wide as is possible.

Some designers are now regarding an amplifier really as a switch for the power supply and a simple trend has been towards heftier and heftier power supplies. A more complicated attitude to power supplies is taken with a new 350 watts per channel Technics amplifier (Fig. 16.12) which, although operating in Class-A, has the voltage rails driven by a separate Class-B amplifier tracking the output signal. Called Class-A +, the advantage is that the efficiency is nearly that of a Class-B amp. Otherwise the size dictated by such a powerful amp operating in Class-A would make it too heavy to lift.

Power supplies are where a lot of development is still to come. Amplifiers have recently appeared that use a power supply which first takes the mains frequency up to around 20kHz and then transforms and rectifies it, a system already in use in television sets. The advantage conferred is that of size; at those frequencies, a very much smaller transformer and smoothing capacitors are required. A new physically small amplifier has also just been reported from the States. Designed by Bob Carver, the designer of the Phase Linear range, it has a power supply based on inductors rather than capacitors. Measuring only a $6\frac{3}{4}$in cube, it, nevertheless, has a 200 watts output.

This is probably as good a place as any to discuss the great debate at present raging about the audible effect of amplifier performance. Particularly in the USA, but also in the UK, a strong movement has sprung up, fostered by the more sensationalist members of the press, which ascribes almost mystical properties to amplifiers. Amplifiers

Fig. 16.12   The heavyweight Technics class A+ amplifier

are described as ripping off or adding ambience to a signal, as making program signal forward or recessed, or as being 'warm' or 'cold' or 'clinical'. A term was even borrowed from the musician's vocabulary, 'musicality', to describe the mysterious property possessed by some amps and not by others. The whole idea was treated with derision by many of the elder and more experienced engineers and, they in turn, were accused by many of the younger – and perhaps more open-minded – engineers of being hidebound. For a while, the audio world polarised into two camps: the subjectivists, whose basic claim was that amplifiers could only be assessed by listening to music signal as the number of parameters affecting performance and the hearing process were too numerous to measure; and the objectivists who believed that amplifier performance could only be assessed accurately by measuring as many parameters as were thought to affect the program audibly. For a while it became quite difficult for the more cynical observers amongst us to stay comfortably sitting on the fence but the last year has seen an acutely intensified interest shown in amplifier behaviour and many of the effects shown by amplifiers can now be explained relatively easily without recourse to mysticism.

It has also become apparent that the relative importance given to amplifier aberrations compared to those of the other links in the hi-fi chain has been much exaggerated, apart from the long term effects which may well be important. Unfortunately, if anyone is doing research into such long term behaviour, it really is too soon at the moment for any valid conclusions to be drawn. Recent extremely carefully controlled listening tests conducted in Canada, the States, and by engineer Martin Colloms in the UK for the magazine *Hi-Fi News & Record Review*, have shown that in the short term, amplifiers which conform to a very high minimum specification cannot reliably be told apart, even by very experienced listeners – provided they are not being asked to deliver more than their designers intended. That last statement holds the key to the whole debate. Some amplifiers, for example, accused of being harsh or gritty in sound in fact are being driven into current limiting with a loudspeaker that presents a very awkward impedance load. For instance, a well known Japanese amplifier offering 75 watts into a 8Ω resistive load will only give around 15 watts at 10kHz into a load similar to that presented by the Quad electrostatic. No wonder the amp sounds uncomfortable, it's probably being driven too hard most of the time, and the designers are at fault for being over-optimistic about their design and the peak/mean ratio of music. But this certainly is not any sort of mysticism!!

What has emerged is the fact that measurements of amplifier behaviour into pure resistive loads, a practice that has been carried out
for years, is unrealistic. Real loudspeakers demand far more of amplifiers than has really been realised and the signs are that amplifier
designs are now taking note. Rough guidelines appear to be:

1 The output should not run into current limiting until driving a
very low impedance indeed, 2Ω or less, so that the amplifier will not
be upset by any of the loudspeakers with which it will be used.

2 THD (total harmonic distortion) for safety's sake should be
around 0·01% or less (although it is a very perceptive listener who
can detect less than 1%).

3 IMD (intermodulation distortion) at supersonic frequencies
brings distortion products down into the audio band and so should be
minimised by having your amplifier linear to quite high frequencies.
(Whether you should use this bandwidth like many Japanese and
American designs – Harman Kardon in particular – or roll off grace-
fully above 20kHz, like many UK amps, is still a hotly debated
subject.)

4 In general, low orders of negative feedback are preferable to high
ones to prevent TID (Transient Intermodulation Distortion) where
the amplifier is unable to control its behaviour quick enough to handle
fast rise time transients without distortion. For a short time at the
beginning of a transient, the full open-loop, perhaps not very linear
response of the amplifier is fed into the output. Also related to the
feedback level and investigated by Finnish engineer Matti Otala, who
first investigated TID, is IIM (Interface Intermodulation Distortion). If an amplifier employs large amounts of negative feedback,
this presents a low impedance path from the output of the amp back
to the beginning and the loudspeaker can influence amplifier behaviour via this pathway.

5 Slew rate limiting is somewhat similar to TID but whereas that
was a function of circuit design, this is a function of the individual
devices employed. It appears that the transistor slew rate should be
fast enough to handle frequencies two or so octaves above the highest
present in the program presented to the input. A young American
designer, Andy Rappaport, has extended this idea to a general
philosophy that the slew rate of a stage should be dictated by the slew
rate of the next stage divided by the gain. This way, no stage is required to do more than it is able.

6   As indicated previously, there are two schools of thought on the bandwidth required of an amplifier: either it should be restricted to the audio bandwidth i.e. 20 Hz to 20kHz, or it should be extended as far as possible. Some recent designs have a response from as near to DC as it is possible to get, right up to somewhere in the medium wave band. The DC end certainly is debatable, after all, no music signal has a DC component, and even the new RIAA recording equalisation curve for disc preamps now recommends that the bass is rolled off below 20Hz. The upper limit, though, is more open to discussion. Recent research has shown that, although it has long been accepted the ear cannot differentiate between HF sine and square waves, the human ear can differentiate between waveforms differing only in their harmonic spectra above the pure sinewave cut-off point (although whether this is due to brain non-linearity or is just spurious; who knows) so it would appear that the amplifier response *should* go above 20kHz. Also from a purely pragmatic point of view, the higher up one preserves the amplifier linear characteristics, the less chance there is of TID or intermodulation distortion.

The trouble is that, like all amplifier parameters, very little research has been carried out into the psycho-acoustics involved – into which parameters are most important and what levels of each are relevant. Up to now, manufactures have played the 'specmanship' game and exploited each new wrinkle ruthlessly as part of the marketing hype. The recent explosion of amplifiers possessing often useless power meters to show the listener if he is hearing clipping distortion or not is a case in point. (Although meters using flashing LEDs provided something to talk about at a recent party!) The public has, of course, gone along with it, but hopefully a major change in attitude will take place. Amplifiers will be designed as a result of the psycho-acoustic research now being carried out, to conform to a certain minimum specification as far as the known distortions are concerned. As I said earlier, why design an amplifier with 0·001% THD when 0·01% is adequate; particularly when even the best cartridge produces THD levels of around 1 to 2%. New distortion mechanisms will doubtless continue to be unearthed but hopefully their relevance to the hearing process will also be sufficiently investigated.

No way has amplifier development reached the end; it's just that designers at the moment are not quite sure which questions to ask and whether the answers they have so far come up with are relevant. The next 10 years should be pretty interesting (Fig. 16.13).

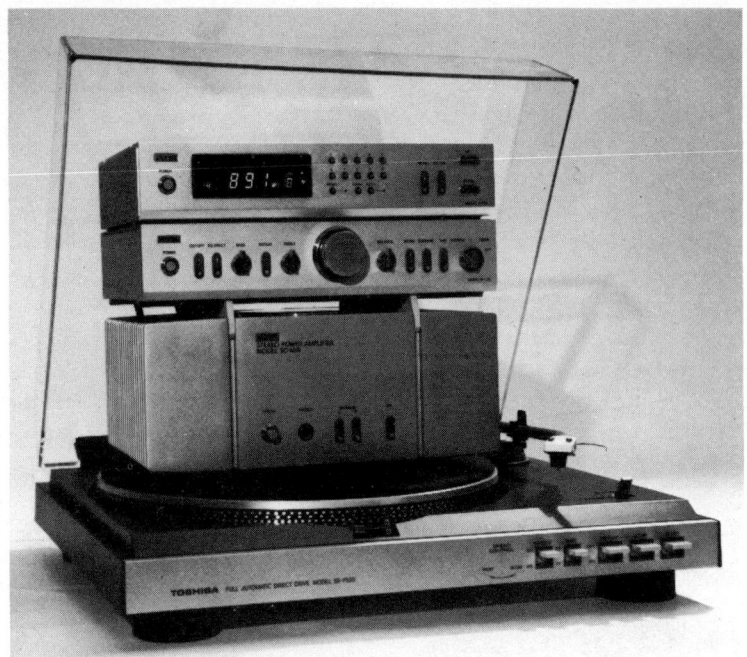

Fig. 16.13 Toshiba's Aurex mini range; a pointer for the future

The disc

A few pages back, I outlined the development of the disc as the primary domestic hi-fi sound source, ending up with the stereo disc in which the two channels are encoded on the walls of a V-shaped groove, each wall at 45° from the vertical. Thus each channel causes the stylus to vibrate with equal vertical and lateral components – a combination of the lateral cut mono record and Edison's vertical cut cylinder. (The vertical cut is inherently more noisy but in this way each channel gets an equal share of this degradation and the two channels have an equivalent performance.) An equal signal in phase in each channel causes pure lateral modulation while equal signals out of phase in each channel cause pure vertical modulation.

This has remained unchanged since the stereo LP was introduced in the 1950s. Changes have taken place in the disc playing system since then, but there has really been only a process of refinement and improvement. The signal is extracted from the disc by a stylus and pickup cartridge combination which converts the mechanical motion into two electrical signals, one for each channel. The cartridge is held above a record either by a pivoted arm, around 9in long, in which case the stylus follows an arc approximating to the straight radius taken by the cutter head, or by a short arm which follows exactly the cutter head path. Because of the more complex engineering required, the latter has never been used widely – there are only three firms in the world making such arms at the moment, Rabco, Revox and B&O (Fig. 16.14) – but it remains the most accurate way of extracting the groove information. The record, of course, has to be spun by a turntable and both turntable and arm are mounted on a plinth system which has to perform two main functions. First, there must be no relative movement between the stylus and the groove as this will introduce a spurious electrical signal. Second, the plinth must isolate the disc playing system from external forces such as airborne or floor borne vibrations from the loudspeakers, and vibrations from the motor driving the turntable.

Fig. 16.14   Revox B790 turntable system with parallel tracking and direct drive

Until fairly recently, most attention has been paid to the pickup cartridge. As the electro-mechanical element in the chain, any aberrations present will have a first order effect on the signal, and so this is where engineers and enthusiasts alike have concentrated. However, it is now generally accepted that poor engineering in the arm and turntable can have as large a deleterious effect.

The cartridge

Relatively few principles have been used to translate the mechanical stylus motion into an electrical analogy, but all are governed by two main mechanical resonances – that of the cantilever suspension compliance and effective mass of the cartridge and arm, in general between 5 and 15Hz; and that of the stylus tip mass and the compliance of the PVC groove wall – in general between 15 and 50kHz. At these resonances, stylus motion is randomised and so doesn't bear much resemblance to the groove information. This can be seen as an increase in crosstalk between channels (Fig. 16.15). Advances have therefore been taking place in placing these resonances out of the audio band, and maintaining the frequency response between them as flat as possible. With the upper response, this is mainly a matter of decreasing the tip mass far enough so that the resonance is at 20kHz or higher. With the low frequency resonance, however, the situation is complicated by the fact that if the resonance is moved to too low a fre-

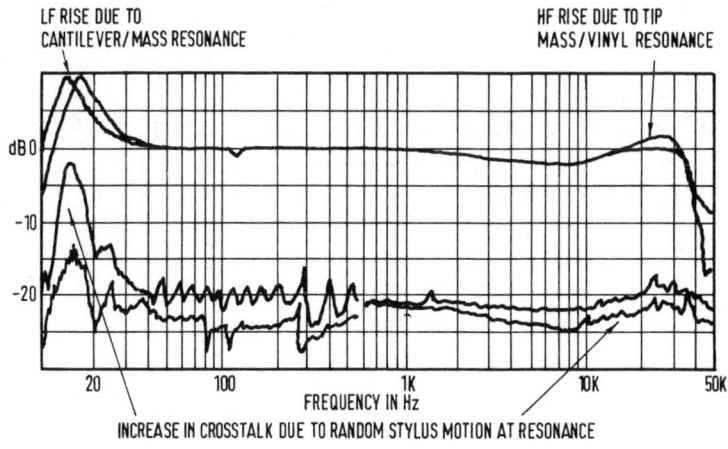

Fig. 16.15 Typical graph of (quite good) cartridge performance

quency, below 9 or 10Hz, it is continually excited by disc warps with deleterious effects on program. If higher than 20Hz, it starts to become audible so we have a limited range available. Unfortunately, one of the parameters involved, arm effective mass, is out of the control of the cartridge manufacturer unless, like B&O, they make both. Thus came a divergence in philosophy in the early sixties concerned with the placing of the LF resonance. (This is rather a simplistic view as other parameters such as trackability – the ability of the stylus to stay in contact with the groove walls no matter how violent the modulation – are involved, but the resonance is perhaps fundamental.)

One group of manufacturers, typified by the Americans Shure, ADC and Stanton, went for very high values of compliance which, in turn, dictated the use of very low mass arms such as the SME. Because of the high compliance, 'give' if you like, the playing force had to be kept to around values of 1gm or so. Other manufacturers, particularly those of the more massy moving coil cartridges (more of which later) and the Japanese went for low compliance which led to the use of high mass arms to keep the LF resonance in the right place. The high mass/ low compliance combination meant higher playing weights to try to retain the trackability.

For a while, high compliance moving magnet (or iron) cartridges, typified by the Shure V15-III where the stylus motion moves a small magnet in front of signal coils, were dominant in the UK but there has been a recent resurgence of interest in the moving coil cartridge. In these, the stylus motion moves two small coils of very fine wire in the field gap of a strong permanent magnet, and some models of moving coil cartridge have offered a noteworthy improvement in sonic performance but avoided the old bugbears of such designs, poor trackability, non-interchangeable styli and over-high mass. Because the coils have to be very small to keep the moving mass low, output and impedance is very low and an expensive step-up device such as a low-noise 'head-amp' or transformer has to be used to match them to the high impedance low sensitivity disc inputs of most amplifiers and this increase in expense has also been a major disadvantage. However, recent models from the Japanese firms of Ultimo and Satin have managed to squeeze up enough output to avoid this extra stage, and the Satins even have user-changeable styli.

A good moving-coil cartridge, such as the Entré-1, Ortofon MC20, Denon DL103D or Supex SD-900 generally scores with lower distortion levels and better crosstalk between channels than their moving magnet or iron rivals, but their complexity of manufacture effectively denies them a mass market. New moving coil designs use air cored coils (Technics), or printed-circuit coils (JVC, Yamaha) (Fig. 16.15) and development is going rapidly ahead in the most critical area, the damping system. Ortofon, for instance, in their new MC30 model cartridge are using two layers of different rubber separated by a platinum washer, and all three layers control cantilever motion at low frequencies, but the softer rubber layer is increasingly decoupled with increasing frequency to provide optimum damping at high frequencies.

Other generating systems are also used in a small way. Micro Acoustics in the USA use a flexed piezo-electric element to derive the signal while a similar principle is employed in the Win Labs strain gauge cartridge. Toshiba in Japan have been hard at work developing electret transducers (an electret has a permanent electric charge in much the same way that a magnet has a permanent magnetic one) and have introduced an electrostatic cartridge. An English engineer, Alec Rangabe, recently responsible for a damping device which attaches to the headshell, has a prototype electrostatic cartridge which tracks at very much below 1gm downforce.

Common to all cartridges, no matter what their generating principle, is the research effort paid to the stylus, aimed at improving contact with the groove walls. This has resulted in a number of patented stylus profiles: Shibata, Pramanik (both named after their inventor), Parabolic, Aliptic, Hyper-elliptical, Fine Line Contact etc., etc., all basically increasing the area in contact with the groove walls at right angles to the direction of motion to get better high frequency performance.

One important development to emerge in 1978 concerned cantilever design. Cantilevers, the short straight bit which supports the stylus and which transmits the stylus motion to the generator, are nearly always a very fine aluminium tube. However, recent work by B&O in Denmark and by Dr Tominari of Ultimo in Japan, has led to the introduction of short cantilevers using a single crystal of sapphire (B&O) or diamond (Ultimo) to give improved pulse performance. Some Japanese manufacturers (Audio Technica, Yamaha, Technics) are using single crystal Boron or Beryllium cantilevers for much the same reasons.

Future trends in cartridge design will deal with improvements in trackability, crosstalk and distortion performance, of course, but the

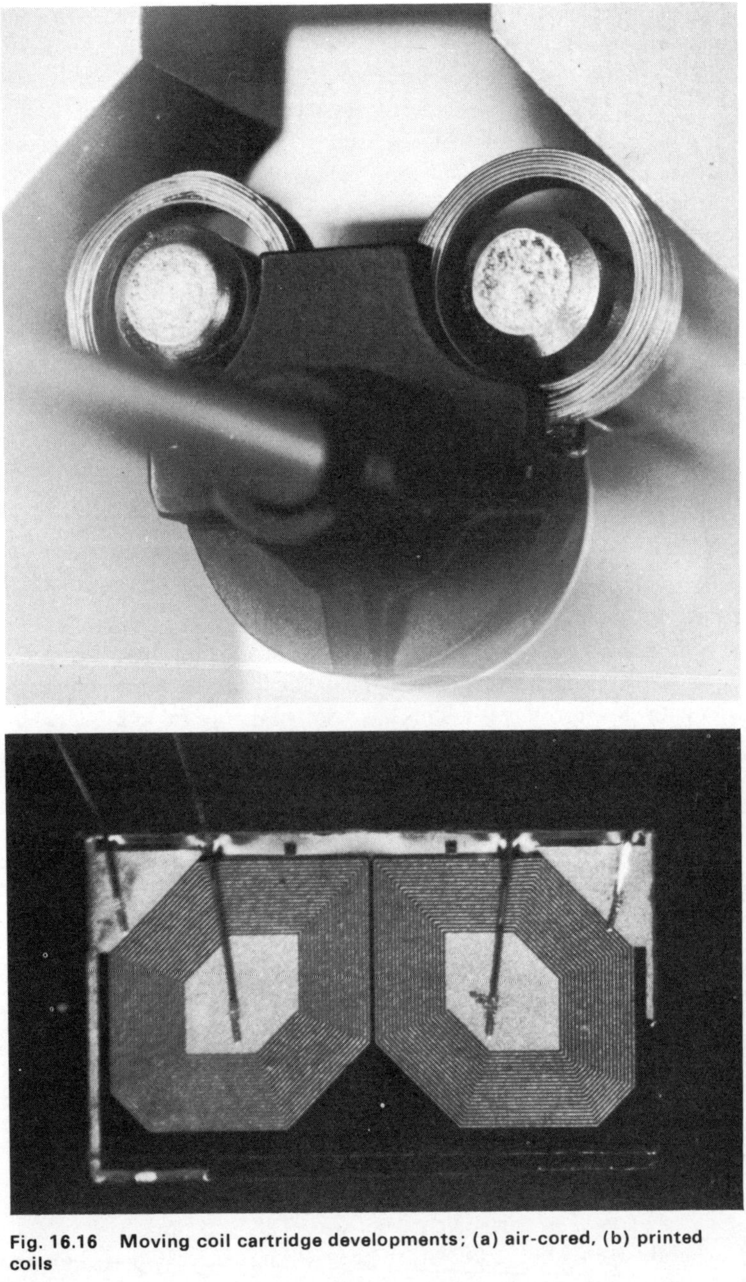

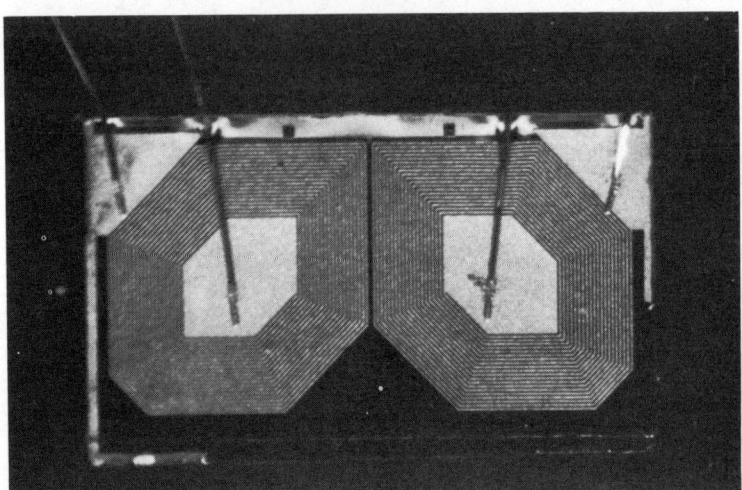

Fig. 16.16   Moving coil cartridge developments; (a) air-cored, (b) printed
coils

main one will be in matching cartridge to tone arm in a less haphazard manner than at present. B&O, Thorens (EMT) and Ortofon are producing cartridges which are meant to be used only in conjunction with a specific tone arm. It could well be that the 'universal' tone arm, never a very satisfying concept, has reached the end of its career. However, in a few years' time all such disc playing hardware will have been made obsolete by the digitally encoded disc, more about which later.

The tone arm

The Americans and Japanese call them tone arms, the British up to now have rather more conservatively called them pickup arms. However, taking into account the very audible effects that the arm can have on the sound of a cartridge, an increasing number of UK engineers are adopting the foreign nomenclature. The arm can have a number of resonances present in the audio band due to complicated modes of vibration and the effects of these have to be minimised, both by choosing the right materials and by using sensible engineering. Tube resonances can be damped: the SME 3009 Series III (Fig. 16.17) has its tube filled with a soft material. Torsional modes are minimised by using very stiff materials for the tube – the SME uses an S-shaped titanium tube with an outer skin of titanium nitride; ADC, Sony and Stax are making much use of carbon-fibre for arm tubes; most companies, though, use aluminium or its alloys and the most expensive arm in the world, the £600 Breuer, made by a retired Swiss watch maker, doesn't use exotic materials. However, it does use well engineered bearings with minimal play and maximum strength in the headshell, for instance, is obtained by machining it from a solid block of aluminium.

Many tone arm defects spring simply from the fact that it is very hard to join together two things without any 'give', interchangeable headshells for instance. ADC do two good arms, one with an interchangeable headshell and one without, and the one with suffers audibly in comparison.

Development will, in general, concentrate on improving those areas outlined, effectively removing tone arm resonances and providing a stable platform for the cartridge, with no spurious movement. As indicated in the section on cartridges, more tone arm/cartridge combinations will appear, designed to complement each other from the start providing, perhaps, mechanical filtering or motional feedback to get rid of the deleterious effects of the low frequency resonance.

279

Fig. 16.17 SME 3009 III arm makes much use of titanium and carbon fibre in its construction

The turntable

The biggest change in turntable design in recent years was the replacement of a high speed motor, geared down to the disc speed by a belt or system of idler wheels, with a motor running at the exact speed required, $33\frac{1}{3}$ or 45rpm, the motor spindle also becoming therefore the record spindle. This *direct drive* principle, for a number of reasons not totally connected with sonic performance, caught on in a big way, particularly with the Japanese. Probably over 90% of hifi turntables these days are direct drive, a large number of them using complex servo control circuitry to maintain the speed as constant as possible. Earlier direct drive decks had a tendency to 'hunt' around the correct speed due to an overlarge inertia in the control system and the so called 'cogging' effect of some of the early synchronous motors, which led to unpleasant flutter distortion but nowadays, good speed control is inherent in nearly all designs (Fig. 16.18).

However, speed control is not the be-all and end-all of good turntable parameters. I remarked earlier that one of the tasks required of a turntable system was to minimise spurious motion between stylus and record groove. Take a direct drive deck and rigidly mount on it an arm/cartridge combination with no unnecessary compliances anywhere in it. Theoretically we now have a superb disc playing system but we have overlooked one important source of spurious motion. It is the absorption of externally generated vibrations and, in particular, those from the loudspeakers. The problem of acoustic isolation seems to have been completely overlooked by a large number of turntable designers who went overboard for direct drive and hence the possibility of good speed control and wow/flutter measurements – a case of applied specmanship without all the specs being taken into account.

You can look at a rigidly coupled turntable, as I described above, as a large expanse of sounding board coupled to a stylus – an acoustic microphone not dissimilar to the horn/diaphragm/stylus used for cutting records before 1927. This microphony was aptly demonstrated on BBC Radio London when they broadcast some speech in the studio being picked up by a stylus resting on a stationary record. With some turntable/arm/cartridge combinations, the level of the signal picked up from the speakers can be only about 20dB below the level of the music signal on the record and this is clearly disastrous!

What is being done to minimise this microphony? One very effective way of isolating the turntable from any external vibration was to mount the turntable, bearing and arm/cartridge on a separate sub-chassis suspended from the plinth by a spring arrangement. The turntable is then belt driven by a synchronous motor attached to the plinth and any motor vibration is absorbed by the belt. Short term speed stability i.e. wow and flutter performance (but not drift) is obtained by using a high inertia platter which acts as a flywheel and a rubber belt machined to very close tolerances, and the record/stylus/arm/platter chain can be made as rigid as you like – noise performance such as rumble is governed really only by the quality of the main bearing, the acoustic isolation being optimised by fine tuning and damping of the sub-chassis suspension. This was the approach pioneered by Acoustic Research, B&O and Thorens in the sixties and the latter still produce superb, moderately priced decks which are unchanged in principle. A very small number of companies have gone for almost no-compromise designs based on the same idea – one of the decks, the expensive Scottish single-speed Linn Sondek LP12 is, amongst audiophiles, perhaps the most highly regarded deck around today. It certainly does offer superb environmental isolation.

Fig. 16.18 High torque capability of the
Technics SP10 direct drive motor which is
being used extensively by broadcasters

If a suspended subchassis is out of the question, then what is the
designer to do? One way out is to isolate the stylus/record from the
platter with a 'lossy' soft mat, and then to isolate the platter and plinth
from the floor and support-borne low frequency noise with springy
feet. This is a design approach much favoured by the Japanese but it
represents a less elegant solution than the elder approach – the soft
mat introduces yet another source of unwanted movement into the
system and the tuning of the feet resonance needs to be carefully controlled. Another way out is to keep the rigid coupling between disc/
stylus/arm but to make the plinth as acoustically inert, as 'dead' as
possible. This has led to the appearance of two new kinds of plinth
material. One consists of mineral particles suspended in a resin plastic

282

and was first used by Trio and Sony in Japan, and has since spread widely. In general, it works well but there do appear to be some problems with energy storage and hence spurious vibrations, just delayed rather than being absorbed and presumably dissipated as heat. The second is used by Lux in Japan and Monitor Audio in England. The plinth is constructed as a sandwich of layers of wood and some very massy material such as lead and this approach seems to offer quite good isolation.

Turntable design is still an area where research into real areas of importance is still going on, and not just research, many companies could do with paying a bit more attention to detail. Many of them, for instance, pay some attention to acoustic isolation, but then put a resonant perspex lid on the deck which effectively negates the effect of all the work done. Eventually, though, it will all be academic when the digital revolution comes. Then all the research carried out on servo controlled direct drive motors *will* have more of a relevant effect.

Quadraphony, etc., etc.

This would seem to be an apt place to examine that white elephant of the 1970s – quadraphony. As anyone who has heard discrete quad from, say, a 4-channel open reel tape recorder would agree, the sound quality can be outstanding and yet, as many hi-fi dealers or record producers would agree, the whole thing is as dead as compressed-air amplification (apart from the Ambisonics work).

What went wrong was in trying to adapt the disc medium to four information channels. There is also a fundamental misconception involved. In stereo, one localises sound sources between full left and right by a mental trick involving the relative intensities of the two stereo channels as perceived by your two ears. (Some people cannot do this at all and cannot be deceived by stereo into hearing virtual images as though they were really there.) With quad, again you have speakers in pairs and image localisation is good at front and back. At the sides, however, this trick won't work and localisation is much less precise. Add to this the haziness inherent in two of the disc systems, CBS's SQ and Sansui's QS (both matrix systems which mix the rear two channels with the two front into the two record channels so that they can be replayed with normal disc playing equipment) and you don't have anything which is going to be too impressive much of the time.

The other systems, JVC's CD-4 and Denon's UD-4, being discrete, offered much better imaging but suffered from needing special disc playing equipment as well as the decoder/amplifier/extra speakers needed by the matrix systems. CD-4, where a supersonic carrier was frequency and amplitude modulated with the rear channels, was probably the most successful worldwide in terms of hardware sales, but the discs were delicate and rear channel performance was easily degraded. EMI issued nearly all its classical discs in the UK for nearly four years in SQ format but this gave centre front images an unacceptably large (to many serious listeners) phase shift, especially when the record was being played in stereo only. CBS and Sansui developed more successful 'logic-controlled' and 'Variomatrix' decoders to help overcome the hardware limitations of the matrix systems, but it was too late; the public had been turned off by the poor results and the resultant indecision of the software people, the record companies, to decide on one universal system. As I mentioned earlier, the only light on the horizon is the work being carried out by the NRDC financed Ambisonics team. They examined surround sound from its first principles – the psychoacoustics of image localisation – and their system has now received the support of the BBC and IBA.

A number of other surround systems are available, deriving the rear channels with varying degrees of success from the normal pair. A technique developed by the American David Hafler, whose success depends on the way in which the recording has been done, simply feeds the difference between the front speakers to a pair at the rear. Much more complex devices use digital delay lines to provide a false ambience signal at the rear, giving a very good illusion of depth, especially when the first echo-time is adjusted to be around the same as that already on the disc. A new device from Advent (Fig. 16.19) in the USA uses microprocessor technology to modify the false ambience signal to conform with that found in existing halls. This could well be a pointer to the future – the record buyer making the decision what concert hall in which to listen to his favourite orchestra without stirring from his chair.

Stereo radio

Stereo broadcasting the world over uses the Zenith-GE encoding system, first used commercially in 1961 in the USA, whereby the main FM signal carries the sum of both left and right channels, thus providing mono compatibility, while the difference signal amplitude modulates a 38kHz supersonic subcarrier (Fig. 16.20). A pre-emphasis

Fig. 16.19 Advent's SoundSpace Controller, a complex reverb unit

(treble lift) is also applied at the transmitter so that when the corresponding de-emphasis is carried out, high frequency noise is decreased. The tuner is instructed to decode the multiplexed stereo broadcast signal by a 19kHz pilot tone. Programme material therefore is sharply rolled off above 15kHz.

Tuner development has progressed steadily on two fronts – improvement on reception and the reconstitution of the original stereo signal; increase in ease of use so that optimum performance becomes automatic without the user having to understand what's going on in order to fine adjust a large number of controls.

Towards the first end, the Japanese in particular have been introducing new circuit ideas at just about every new exhibition. Pilot-tone cancelling circuitry, improved filters such as the SAW (surface acoustic wave) filter introduced by Technics and JVC, phase locked loop decoder circuitry, new techniques such as Trio's pulse count

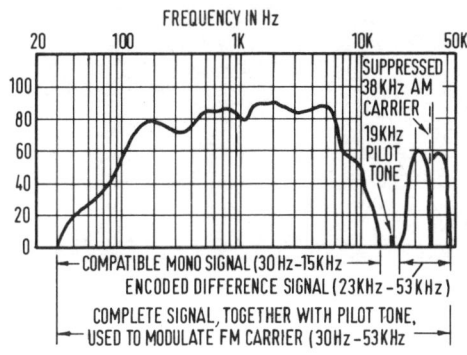

Fig. 16.20 A multiplexed FM stereo signal

decoder, automatic frequency control which turns off when a hand touches the tuning knob, etc., have all led to dramatic improvements in signal to noise ratio, capture ratio (the measure how the tuner rejects an unwanted station on the same or slightly different frequency), distortion caused by tuning drift, etc. Tuners being introduced by Yamaha and Aiwa select automatically the optimum selectivity required for minimum distortion.

The most obvious change to happen with tuner design is the adoption of quartz crystal control for the local oscillator, enabling the tuner to stay very accurately tuned to the centre frequency around which the signal is modulated, thus minimising distortion. An added bonus of this is that it becomes relatively easy to replace the bulky analogue tuning scale with a digital LED readout of frequency, leading to very neat unobtrusive design typified by new models from Technics (Fig. 16.21) and JVC. One big disadvantage, however, is that with such frequency synthesised tuning, it is more convenient to tune in discrete frequency steps and whereas the Japanese have, in general, opted for 100kHz steps, some European stations have frequencies which are a multiple of 25kHz. In that case the 25kHz off-tune condition would introduce a lot of distortion and negate all the advantages gained. This situation is in a state of flux at the moment, but some tuners are being introduced capable of coping with 25kHz steps.

This, of course, is a case of fundamental circuit design affecting the ergonomics involved and much work is going on to make tuners easier to use. Going hand in hand with the introduction of quartz synthesis and digital tuning has been the introduction of microprocessors to control such functions as station selection with the minimum of inter-

Fig. 16.21 Tiny Technics tuner and matching microprocessor unit

ference from the user. Such tuners as the Revox B760 can have up to 15 stations preset in the tuner memory (kept active by battery back-up when the tuner is turned off) and a Technics microprocessor add-on unit enables the tuner to be preset for up to 24 operations including turning itself on, tuning in to a preset station, rotating the aerial for the strongest signal, turning on a tape recorder, recording a programme, and then turning everything off at the end of the programme – all that needs to be added is a few more ROMs, RAMs, EPROMs, etc. and the device could be programmed to enjoy the music!

One experiment which has been tried widely in the USA but only by two IBA stations (Capital Radio and BRMB) in the UK, is the application of the Dolby-B noise reduction system to FM stereo broadcasting. Some degree of compatibility with non-Dolby receivers is given by the adoption of less pre-emphasis but it appears that the benefits are not great enough at the moment to persuade UK stations to adopt Dolby broadcasting full time. The same applies to the experiments carried out on quad broadcasting – the attitude everywhere seems to be one of wait-and-see.

Tape recording

Since the wholesale revolution in sound recording caused by the introduction of the tape recorder, improvements have taken place in two separate areas – improving the tape coating to give better frequency response and distortion performance, and introducing new hardware formats. The most revolutionary change in the second category was the introduction of the Philips 'Compact cassette' some 15 years ago, with $\frac{1}{8}$in tape running at $1\frac{7}{8}$ins. Engineers laughed at the time, cassette performance was generally in the area between 'awful' and 'are you sure this is me speaking?' but in the last five years, improvements in tape coatings and tape head technology and the addition of noise reduction systems such as Dolby-B, Telcom, super ARNS and Toshiba's new ADRES have led to such improvement that the average domestic cassette machine is capable of better performance than the domestic open reel machines around at the time of the cassette's introduction. There are now even three head cassette machines around, with the record and replay head gaps optimised for their unique function.

Tape coatings progressed from ferric oxide, via chromium dioxide, back to ferric oxide again, either alone, with special care taken on crystal shape and orientation; or doped with other ferro-magnetic elements such as cobalt; or having a ferric oxide layer coated with a very thin layer of chromium dioxide. With all these different coatings available to the consumer, it has been almost mandatory for machines to now appear having fine control of bias current in order to get optimum performance. What is more unusual is a new generation of machines which adjust their bias automatically, using of course microprocessors, which have wheedled their insidious way into cassette technology. A less serious use of them occurs in Optonica machines where you can program the machine to play back the tracks on a pre-recorded cassette in any order the user wishes (Fig. 16.22). A less frivolous advance is in the field of metering. For years, tape machines of all kinds have used average level meters of the VU kind which have proved to be next to useless for accurate monitoring of recording level. A first attempt to improve meter performance involved the addition of a peak reading LED, but in the last year three different types of new meters have appeared. Sony are using a liquid crystal meter (Fig. 16.23) with the level indicated as a bar graph. Technics are using an incandescent display bar graph while JVC are splitting the audio spectrum into five separate bands and indicating the level of each with LED bar graphs. The advantage of such displays is that their intrinsic mechanical inertia is effectively zero so it becomes an easy task to tailor electrically the ballistics required whether VU or peak reading.

Cassettes knocked the bottom out of the cheap end open-reel market but the 'serious' end, the open-reel market, is extremely healthy with Revox, Uher, Tandberg and Ferrograph in Europe holding their own against very strong Japanese competition from Teac, Pioneer Sony, Akai, and in particular, Technics whose RS-1500 machine (Fig. 16.24), with a transport using three of the direct drive motors used in

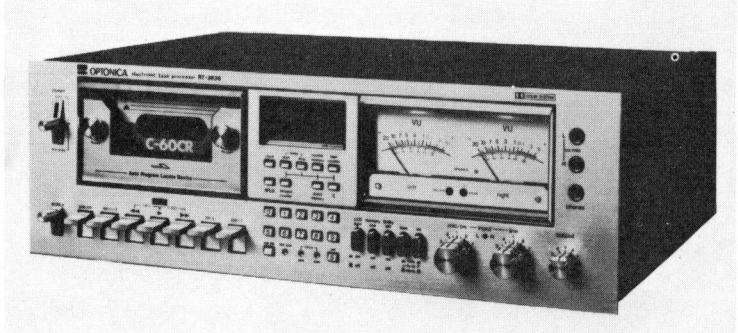

Fig. 16.22 Optonica microprocessor cassette deck

Fig. 16.23   Another pointer to the future – bargraph metering for tape decks

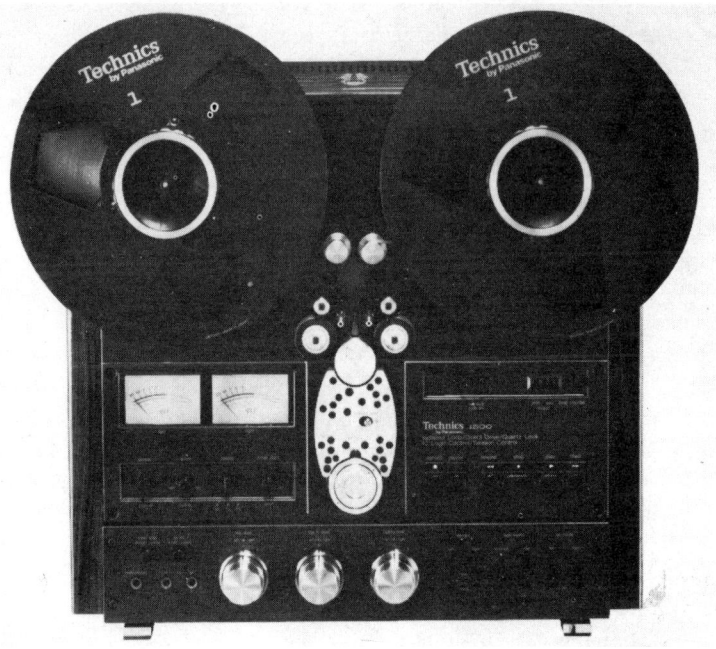

Fig. 16.24   High level of technology in a Technics consumer tape machine

their professional SP10 turntable, is the forerunner of a family of machines from $\frac{1}{4}$ track to 24-track.

Toted in 1975 as a more convenient replacement for open reel and supported by machines from Sony, Teac and Technics, Elcaset proved to be a typical Japanese answer to a problem no one had ever realised existed. Although scoring above the compact cassette in every way with double the speed and track width and the facility for editing, it arrived 10 years too late for its place in the marketplace, and coupled with very highly priced software, the public just didn't bite.

Two major developments loom on the tape horizon – the first is digital recording which I'll cover in depth in the next section; the second is the introduction of iron powder tape. The use of metallic iron is rather than its oxide, with its much higher level of magnetisation, gives performance from cassette comparable with that from an open-reel machine using oxide tape at 15in/s. Iron tape is hard to manufacture – powdered iron makes a good explosive – and the high current required presents severe problems to the designers of erase and record heads as normal heads saturate at that current level. However, all cassette machine manufacturers have models ready to take advantage of the new tape but at the time of writing (January 1979), there is no international agreement on suitable bias and equalisation. When these are agreed, two-speed machines will appear, using $1\frac{7}{8}$in/s for serious use and $\frac{15}{16}$in/s for normal domestic use.

The digital revolution

Much of the controversy raging on various components in the hi-fi chain can be regarded as the last fling of a soon-to-be obsolete technology. The treatment of audio signals in purely analogue terms will very soon be replaced with a whole new way of doing things based on the digital technology spawned by the computer industry. In the recording industry, the revolution has already occurred, with several companies, Denon in Japan, Enigma and Decca in the UK, Telarc and Soundstream in the USA, issuing records made with modified video recorders. PCM (pulse code modulation) (Fig. 16.25) recording has one large advantage over its analogue equivalent. In PCM, the amplitude of the audio signal is examined a very large number of times a second, each sample being then described as a binary digital 'word', made up of 'bits'. Each bit can be at high level or a low one, and obviously the more bits – the longer the word – the more

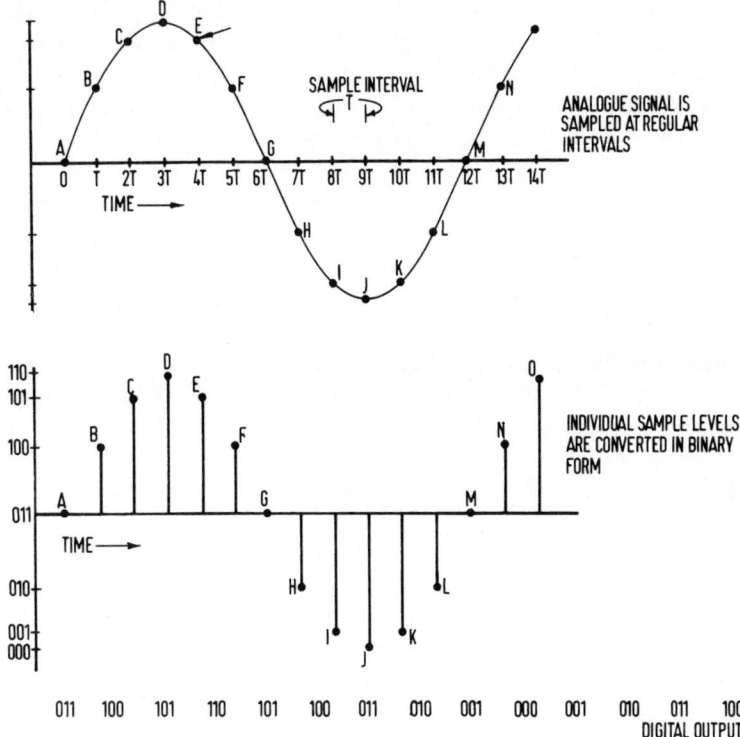

Fig. 16.25   Pulse code modulation

accurately the signal can be described. When applied to tape recording, all that is required of the tape process is to be able to record and then distinguish between just two magnetisation levels instead of the infinite number required of an analogue process. Thus any imperfections in tape or recorder have no effect on the signal, all the improvements that designers have been sweating over are – bar one – irrelevant. This also means that copying will not introduce any degradation and the process brings the listener that much nearer the original event. However that one exception I mentioned above is extremely important. In order to describe accurately all the frequencies present in an audio signal, you need to examine it at least twice as many times as the highest frequency present. For the audio bandwidth to extend to 20kHz, we need a 'clock' frequency of at least 40kHz. (If the clock frequency is below the highest present in the signal, a particularly unpleasant form of distortion called 'aliasing' occurs.) However, coupled with our 40,000 words every second, each word has itself to consist of many bits – at least 13 – giving a required bandwidth of at least 500kHz per channel. This is impossible with the type of tape machine normally used but is perfectly feasible with the rotating-head technology developed for video recording. Technics, JVC and Sony have introduced PCM adaptors for their ranges of video cassette and open-reel recorders to bring such sound recording qualities within reach of the consumer.

One large hang-up still exists, however – standardisation. The advantage of analogue recording at the moment is that you can record your orchestra or whatever in New York, overdub the singers in London, re-do some of the trickier bits in Paris and then use that tape to cut a record in Tokyo. Digital recording unfortunately is still in a Tower-of-Babel situation with every company involved insisting that their's is the only true number of bits, tape format, etc., etc. Until an international standard is agreed on (and fortunately everyone remembers the death blow dealt to quadraphony by much the same kind of wrangling) digitally recorded records will stay out of the mass market and remain in the province of the small enthusiast company.

Much the same criteria apply to the development of a digital disc system. Several competing designs now exist and all of them do away with the antiquated amplitude modulated groove/stylus/tone arm system with all its inherent associated problems. Much use is made of laser beams reading digitally encoded signals on a glass disc surface, turntables which change their speed of rotation to keep the angular speed constant etc., and estimated prices for digital disc players are not all that above a normal high quality analogue system. The improvement in quality is astounding – if the original performance is *first generation* then a conventional record is normally *6th or 7th generation*, a digital disc is effectively *2nd generation*! All these disc systems are at the moment totally incompatible and while this state prevails, there will be no way in which the people supplying the essential software, the *records*, will take advantage of all the benefits. Again, everyone concerned remembers quadraphony.

As a taste of things to come, in Spring 1979 Philips demonstrated its Compact disc player using a 4½inch diameter optically read

Fig. 16.26 Philips Compact disc player using 4½inch optically read digital disc that plays in stereo for one hour with over 90dB signal-to-noise ratio

reflective disc (Fig. 16.26).

Assuming, though, that these problems can be resolved, hi-fi in the eighties will be of a quality thought unbelievable in the fifties when so much of the present technology was in its infancy. Looking forward to the turn of the century, we could well see a convergence between hi-fi and video technology. Already the video and digital audio disc players are similar enough that only one player will be required for both – the next step will be to have just one transmission system. The form this could take is anybody's guess – my own idea is that small sound transducer elements could be interspersed with the colour elements of a flat large screen video, giving in effect a wall of sound. These elements could be matrixed to give all the stereo information required – and not just stereo; Technics and JVC in Japan have done some very successful work on producing surround-sound from speakers in the front plane only – it is quite uncanny to hear a sound coming from behind when the speakers are in front of you. The future will be very interesting indeed!

17   In-Car Entertainment

Donald Aldous

A dozen or so years ago, in-car entertainment (ICE) often meant placing a portable radio set on the back seat, with a little singing or whistling to oneself thrown in for good measure. Of course, one must remember that the original Philips car radio dates back to the thirties but these designs were bulky, needing robust support back and front, and put a heavy load on the battery.

Today the motorist has available a bewildering variety of car radios (AM, FM, mono, stereo), cassette/cartridge players, stereo tape players, in various formats, sizes and combined with radio receivers of high performance (Fig. 17.1).

Transistor circuitry has reduced the size of ICE systems, and the drain on the battery is no longer a problem even when the car is stationary in traffic jams. The typical frustrations of driving from traffic lights to road works and wandering pedestrians, can be alleviated by a properly installed car radio/tape system. The radio can advise of traffic hold-ups and how to avoid them, inform on weather conditions, and keep us in touch with the latest news. Hi-fi enthusiasts may regard in-car stereo as just a gimmick, but playing cassettes (Fig.

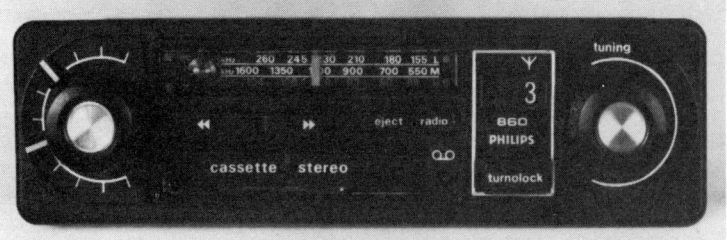

Fig. 17.1   Philips AC860 AM/FM radio with IAC, and stereo tape player

Fig. 17.2 Blaupunkt Marburg CR motorises station preselect FM/AM radio with stereo cassette player from Bosch

17.2) or cartridges of one's own choice is often the only opportunity some busy people ever get to hear their favourite works uninterrupted. Broadcast music programmes are another fruitful source of mobile music.

To get a grasp of the car radio/tape player growth pattern in four European countries, Fig. 17.3 shows a chart prepared by Grundig with the 1979 trend estimated.

Fig. 17.3   Car radio/tape player market

Country	1977	1978	1979
France	14%	17%	22%
UK	23%	26%	29%
Italy	12%	14%	16%
Germany	49%	57%	74%

Biased as they may well be, many ICE system manufacturers look upon a radio or stereo, or preferably a combined unit, as a worthwhile investment in a car because it enhances the vehicle's value, improves the environment in which one drives, and helps to make a better and safer driver. The last point could be regarded as controversial as, under some road conditions and speeds, any music or speech within the car might be a distraction rather than a soothing influence.

Most major towns have a car radio/tape centre specialising in supplying and fitting ICE systems suited to one's car and pocket. Nevertheless the array of equipment is daunting and in the bigger cities some dealers will concentrate on one or two makes but carrying all the models within the brand. The spectrum of prices is also very wide from about £30 for a simple LW/AM car radio to the top-end Blaupunkt Berlin system at £740 plus VAT.

The Berlin model consists of three units: a small remote control head mounted on a flexible arm, a radio receiver (FM mono/stereo, MW, SW, LW) and a stereo cassette player/recorder. The remote control head provides automatic station search and all functions are controlled by touch-buttons on the head's sides and front. The cassette unit fits into the radio aperture on the fascia while the radio section can be mounted anywhere in the car.

Another advanced design is the Hitachi CSK 501 Digital model. This is a combination cassette player with LW, MW and FM stereo radio featuring digital read out, microprocessor memory tuning and FM noise limiter. Physically, even with all these features, it fits into any standard (DIN type) dashboard space. The microprocessor records and remembers the exact wavebands of six preselected radio stations – three stereo FM, two MW and one LW. Tuning requires just a fingertip touch on a preset button. For a wider selection, the CSK 501 model incorporates a self-seeking device that shifts automatically from one station to the next at a touch of a button. Whichever band is chosen, it is identified in red figures on the digital frequency indicator which has an automatic dimmer call for LED output. In addition to these facilities, as well as a first class cassette cradle – a car environment needs rugged construction – it has a wide range variable tone control and a pull rotary loudspeaker balance.

For the affluent car radio/audio enthusiast, National Panasonic have their Car Component Range (Fig. 17.4) which includes such hi-fi features as separate bass/treble tone controls, front-to-rear fader controls for 4-speaker system, preamplifier and a stereo power amplifier plus an advanced cassette deck with Dolby noise reduction on the D11 version.

Philips have recently introduced their Hi-Q car radio systems and accessories. One Philips development is the SDS/SDR circuitry which

Fig. 17.4 National Panasonic Car
Compo with stereo cassette, stereo
FM/AM tuner, stereo preamplifier
with full eq, and separate power
amplifier

automatically adjusts both channel separation and HF noise suppression to match the signal strength. Automatic stereo-mono switch-
over is common, but such a direct change from mono to stereo
normally causes a drop of around 20dB in signal-to-noise ratio,
accompanied by a 'plop'. The latest Philips integrated circuit, Sliding
Stereo/Signal Dependent Response keeps the S/N ratio steady and
the changeover imperceptible (Fig. 17.5) by progressively reducing
high frequencies as the incoming signal strength drops, and thus the
more offensive noise.

Coming down from these dizzy heights, the do-it-yourself enthusiast
may be tempted to fit the car radio/tape system himself. If you want
to obtain optimum results, this job does require some know how. Of

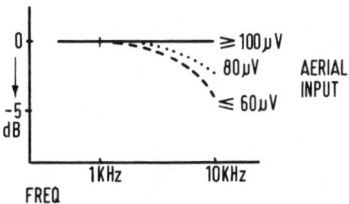

Fig. 17.5 Philips SDS/SDR signal dependent response system which reduces
high frequencies as aerial level reduces

course any standard of installation will get the system working, but possibly far below its peak performance. In the long run, a wise move is to have the system installed by a specialist dealer and this usually does not mean the local garage service department!

On the assumption that one's car is not already fitted with an AM radio unit (Fig. 17.6), AM/FM unit, cassette or cartridge tape player, or one of the several combined tape player and radio systems, the first decision has to be whether to buy an AM unit (usually covering medium and long waves) or an FM/VHF unit. Certainly never buy a cheap unknown make of FM receiver as your outlay is better expended on a good AM unit. Add-on FM units were available, but it is advisable to have an AM band as well, although FM/VHF coverage is now improving. Again a car radio with stereo capability is not a useless luxury as, apart from BBC stereo transmissions, local commercial IBA stations around the UK radiate programmes in stereo. These are receivable though only over a limited area.

The superior sound quality obtainable from FM broadcasts will be known to any hi-fi devotee, and even within the confines of a modestly sized car, with carefully fitted loudspeakers, the improvement over the average AM signal is immediately apparent. This VHF/FM method also offers minimal fading under bridges, but the disadvantages include interference suppressions problems, which can be tackled by Blaupunkt ASU or Philips IAC circuits, of which more later. The other factor, of course, is the comparatively high price of good units against simple AM sets.

Turning to the AM bands, we have reliable reception over a wide area, fewer interference snags and cheaper units available. On November 23 1978, the BBC wavelength changes came into force with Radio 4 transferred to 1,500 metres (200kHz) although the VHF/

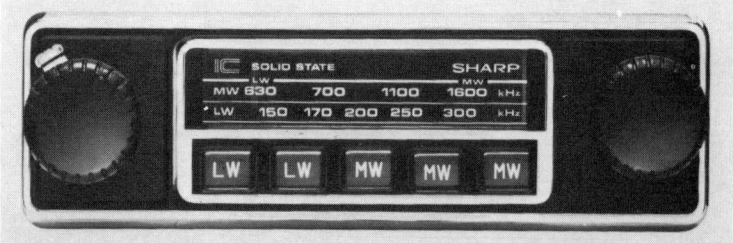

Fig. 17.6 Sharp AR947 AM radio with MW/LW and five preselectors

FM frequencies are unaltered, and even the least expensive car radios include an LW band.

Looking at the demerits of AM, the sound quality is limited, some fading under bridges or near tall buildings, overcrowded bands, and no stereo capability.

Circuitry advances

The complex Blaupunkt Berlin system and the Hitachi Digital model have already been mentioned and progress in refining car radio receiver circuitry started with the introduction of the transistor. Fitting semiconductors in the output stage made obsolete the vibrator power pack with its bulky components designed to produce the high voltage required by valves.

Early valve designs were fitted with 12V valves in the control unit, having dimensions for these manually tuned models of 7in × 7in × 2in. The later press button models included an additional amplifier circuit, leading on to all transistor radios fitting fascia slots having the same overall measurements. It must be mentioned that today these dimensions seem to vary somewhat according to the type and make of unit, but are usually suitable for in- or under-dash fittings by the use of gimbals or brackets. An alternative method of fitting is a radio console unit for installing over the gearbox – these will house both radio and loudspeaker.

The current consumption drain on the car battery has now dropped from about 2·5A of the early vibrator valve power systems, to less than 1A of the modern all transistor design.

Chassis polarity

An important point to grasp, both in regard to radios and tape players, is the polarity of the unit in relation to the particular car. Until about 1968, all British cars were fitted with positive earth electrical systems but towards the end of the sixties, British car manufacturers agreed to convert to negative earth, as used on the continent.

The reason for this convention being important is that one battery terminal is earthed to the bodywork so that only one power lead need be routed to the various circuits with the return path via the car metal work. To avoid destroying the transistors, it is essential that the same pole is earthed as with the car battery. Today the bulk of car systems are designed for negative earth, although some models are still on the market with dual polarity (the change is made by moving over an internal switch or resoldering links across a terminal strip). Polarity converters for non-convertible equipment incorporate ingenious circuitry. The direct current input is first changed to AC which is applied to a 1:1 transformer and this output is rectified and smoothed down to DC again. Within the unit, neither pole of the output is earthed and both remain floating. This arrangement provides positive or negative earth, as required, irrespective of the input polarity.

Car aerials

Specialist car radio installers all seem to agree that a car system is only as good as its aerial. Philips engineers maintain that a figure as high as 80% of service and performance problems can be attributed to inferior or badly fitted aerials. It has been found that many aerials are totally unsuited to FM requirements.

Radio waves cannot penetrate the sheet metal bodywork of a car, being screened by the Faraday Cage principle from external transmitted electrical fields, so for adequate reception of radio signals an antenna mounted outside the bodywork is essential. The most efficient designs have evolved into the conventional rod antenna, although this is by no means a standardised product with different models designed by leading manufacturers around the world.

Obviously not every aerial is suitable for all cars and receivers and some designs concentrate on style. Each car may need different mounting facilities and an aerial may be retractable as well as being electrically operated. A few manufacturers have discontinued making electrically operated aerials on grounds of reliability and limited sales, not to mention vandalism of these expensive types. In addition to these motorised aerials which extend when the car radio is switched on and automatically retracts on switching off, a recent development is the electronic type car antenna. This has a very short stub which is fully retractable or removable, and the resultant loss of signal voltage is compensated by an amplifier in the base of the unit. Specialist dealers have lists of aerial types suitable for all makes of car. A trend today is to have the aerial fitted when the car is built. The manufacturer for most of these aerials in this country is Antiference Ltd.

In view of the effects of weather and constant 'up and down' movements, long term reliability plays a decisive part in choosing an aerial. Telescopic rods can be made of stainless steel, or nickel or chrome plated brass. To resist corrosion, in the best designs of antenna all brass components are heavily nickel and chrome plated. Water ingress must be avoided by the use of special die-casting and injection moulding techniques. Good contact must be assured, even at high driving speeds, and the masts must withstand up to fivefold bending forces. Constant vibration must not disturb the connection between the base of the aerial and the lead-in cable to the car radio.

Retractable and disappearing aerials are the most popular types, usually fabricated in four to seven sections with overall lengths of about four feet as an average. This works out about the length of a quarter-wave antenna at the frequencies in the VHF/FM band around 93MHz, in which we are interested.

Another group of car aerials includes screw-on steel whip types, window models and collapsible types that fold down when meeting an obstacle. The newest type is an aerial integrated into the windscreen. It consists of a fine silver conductor embedded in or printed upon the glass and tuned to provide 'peak' reception. This device has the virtue of being protected from dirt and damage, but it can only be supplied by the windscreen manufacturer. On the Continent, we understand, to meet the wishes of mobile TV addicts, television aerials are available. These are 25in dipoles and are fitted on the rear wing. With one or two of the mini-TV receivers now available in the UK, these units could find a following here, but we hope only for the back seat passengers!

For the DIY enthusiast economising on his installation, coat hangers are occasionally used for car aerials. Certainly these can provide limited reception, but usually only of high signal strength local stations with no guarantee of reliable pickup. Not recommended.

For optimum performance of a car radio, the aerial system should have a capacitance of between 65 and 75pF (a European standard) and needs to be less than 100pF. Values much higher than this figure are usually outside the range of the compensating trimmer to correct. Incidentally, many Japanese and American car radios are designed for aerial capacity of 95 to 100pF as are certain Japanese aerials. Obviously when these are used with European units, a reduction in output must be expected.

Television coaxial feeder type cable is not suitable for coupling car radios to the aerial system as it has too high a capacitance for this special purpose. The correct type is a low capacity, copper braided air-spaced coaxial cable with a universal fitting plug. If it is necessary to extend the feeder cable, remembering the inbuilt capacitor compensation; an aerial extension lead must never be shortened nor for that matter must the actual feeder cable be cut – simply 'lose' it along the length of the car by snaking it, not coiling it.

Interference suppression

Tape playing equipment installed in cars require no suppression but it is essential for radio systems. There are two basic forms of interference, radiated and conducted, and before trying to discover the source of interference it is necessary to find how it is getting into the radio receiver. Components used to tackle this annoying intrusion are capacitors fitted between equipment supply lead and earth and inductors or chokes fitted in series with the item being suppressed, some having a core of ferrite material.

Resistors fitted in the HT ignition circuit limit the current flow, and interference pulses are correspondingly reduced in amplitude. A principal source of interference is the distributor and a resistive suppressor (of about $10k\Omega$ in value) should be fitted in series with the lead between the coil and the distributor centre HT connection. It is a legal requirement for new cars to have distributor suppression incorporated so this fitment will only be needed on older vehicles. Screening by putting a metal case or sheath around the unit or component and then earthing the case, has an obvious application.

Fibreglass body cars need perforated screening foil for suppression purposes. Incidentally, fibreglass body cars can basically be considered as an engine and transmission with wheels, on a metal chassis. The body serves to keep out the elements, but has no screening properties. For an interference point of view, it is necessary to find a good earth point to attach the braid to the radio receiver.

Philips has developed an Interference Absorption Circuit (IAC) to help in eliminating many types of HF interference, spark noises, etc. heard on FM derived from the car electronics, nearby vehicles and other external influences. The IAC circuit can be regarded as an electronic switch which operates the moment interference occurs. The IAC is inserted between the FM demodulator and audio amplifier, but is switched off on AM.

The audio signal from the FM detector with its spikes of interference, is passed both to a delay circuit and a high rise filter simultaneously. The high rise filter removes the audio and passes on the interference spike which is fed to a pulse converter and emits a key pulse. This key pulse triggers a gate (electronic switch). At the end of the interference and subsequent key pulse, the gate closes and allows an audio signal to go through the delay circuit to the audio amplifier. The delay circuit is used to counteract the delay in producing key pulses, due to longer circuit length, and so the gate is opened or closed in synchronisation with interference. This circuit handles peaks, so with very bad cases of ignition interference some spikes and noise will remain (Fig. 17.7).

Blaupunkt tackles the problem of interference on the FM waveband with an AUS (Automatic Suppression Unit) circuit fitted to all their FM mono and stereo receivers. It identifies intrusive signals from the car's ignition and minimises this undesired interference.

Ranges of suppression components are produced in packeted formats from various sources. For example Philips market a batch of such components including 1 and 2μF parallel capacitors with various terminals suitable for connection to most vehicles. Other items include feed-through capacitors (2·5μF and 0·5μF), resistive suppressors for fitting in distributor caps; screened plug-cap suppressors plus an inductive suppression unit; supply lead filter (housing inductive and capacitive components) and bonnet springs which are useful for earthing the rear edge of forward opening bonnets.

Tape players

Fig. 17.8 reveals the differences between the two enclosed tape systems fitted in cars and in the home, if we ignore the open reel tape machines.

Philips pioneered the Compact cassette system, introducing the tiny portable 3300 battery recorder in 1963 and in one step solved the stereo/mono compatibility problem by recording all their Musicassettes in $\frac{1}{4}$-track stereo, but with the left and right signals on adjoining tracks (1:2 and 4:3) instead of the separated 1:3 and 4:2 tracks used on conventional 4-track open reel machines.

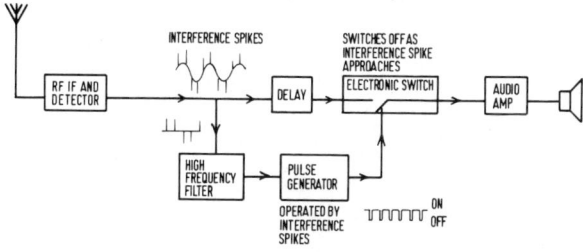

Fig. 17.7   Philips interference absorption circuit for VHF/FM radio

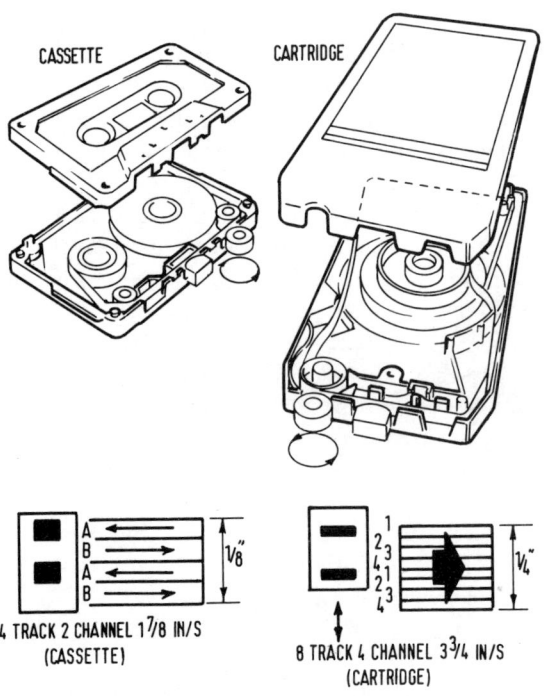

CASSETTE   CARTRIDGE

4 TRACK 2 CHANNEL 1⅞ IN/S
(CASSETTE)

8 TRACK 4 CHANNEL 3¾ IN/S
(CARTRIDGE)

Fig. 17.8 **Exploded views of tape cassette and cartridge**

Competing with the Compact cassette system, although aimed mainly at the automobile market when launched in the USA, is the 8-track continuous play stereo cartridge divided into four pairs of tracks, with each coupling representing a left and right stereo programme. This method was initiated by the Lear Jet Corporation in 1965 and later employed by Motorola, RCA and the Ford Motor Company for their vehicles.

For completeness the discrete quadraphonic 8-track cartridges must be mentioned; each of the four separate sound tracks is played through its own loudspeaker, two in the front and two in the rear of the car. These cartridges are compatible and so normal 8-track cartridges can be reproduced on the 'quad' player. The Radiomobile Stereo 4 Model 109QS is one example in this category. Yet another type of player was available termed a 'matrix' unit which electronically divided the signals from a standard 8-track stereo cartridge into four sources feeding the sounds to four speakers.

As innovators of the Compact cassette tape system, Philips believe that this neat spool-to-spool device packed into a small plastic case, against the larger endless loop continuous play cartridge system, is superior. In practice each method has its advantages and disadvantages. For instance, cassette hardware is obviously more compact than equipment for handling cartridges. Cartridges were intended as a playback system against the cassette concept of a record/replay system.

Cartridges run at twice the speed ($3\frac{3}{4}$in/s) of cassettes using standard $\frac{1}{4}$in wide tape and potentially this means that they have wider frequency range and lower background hiss, but on the other hand have no fast wind facility or reverse wind. Making one's own recordings on cartridges is not normally possible, although this is easy with the cassette.

Cartridges can change tracks in an awkward position in mid-programme while some car cassette models permit recording from microphone or the car radio direct. Tape cassette life is usually longer than that of a cartridge, but the former need to be turned over by hand at the end of a run unless an auto-reverse unit is fitted.

From this survey, it can be seen that each format has its merits and probably the choice for a car system will be determined by the tape method installed at home to allow interchangeability, linked to the extent of pre-recorder cassettes available. Today the tape cartridge must be considered to be in decline as recent improvements in the cassette format, including refined tapes, better tape electronics and drive mechanisms and compatibility with domestic systems, have placed the latest cassette models in the forefront of ICE systems. Tape cartridges can, of course, still be bought and many older cars are fitted with this format.

Combination units – that is radio and tape players in one housing – are becoming increasingly popular from such well known brands as Blaupunkt (Bosch), Philips, Hitachi, Sharp (Fig. 17.9), National Panasonic and Clarion. If you seek (and can afford) the features and sound quality of a home stereo system, these are now available in a condensed format at a price. For example the Motorola TC887AX AM/FM stereo cassette player can be bought with a matching EQB-3000 stereo graphic equaliser booster. The booster unit can be fitted underdash and the leads from the booster take their signal from the speaker outputs of the associated radio. The EQB-3000 front panel

Fig. 17.9 Sharp RG5750 stereo cassette player with stereo AM/FM radio covering LW/MW/VHF

has a power switch which, when in the off position, provides direct feedthrough from radio to loudspeakers, but while on the five equaliser lever controls takes over from the main radio and includes power output indicating LEDs calibrated to flash or light for output levels of 1, 2, 5, 10 and 30W combined power of both channels.

The mind does boggle a little at the thought of 60W dissipated within the confines of a motor car, but Sony offer such a model and the Jensen car range has its system R430 which incorporates a completely separate boot mounted stereo power amplifier as well as a twin amplifier section in the main unit. The R430's bi-amp selectively sends HF power to the front speakers and LF power from the main amplifier to the rear speakers. Other features include electronic switches, an automatic tape alarm and loudness and muting controls are fitted. Another US car high power system has the name Mind-blower.

Domestic tape equipment these days incorporates dynamic noise limiting circuits (such as Dolby-B) to improve overall performance, and this Jensen (as does the Pioneer, and a few other makes) include this circuitry in car units. However prerecorded Dolby-B music tapes played back without Dolby 'decoding' provides a degree of automatic compensation for a noisy listening environment and thus can be helpful in a moving car.

As to the future, Philips demonstrated its new Compact disc system in March 1979. This uses a $4\frac{1}{2}$in diameter digitally encoded disc, optically played, that is just the size to fit in a DIN dashboard cut-out. With a 60 minute playing time and quality exceeding anything currently available, this is expected to replace Compact cassette in cars during the eighties.

Loudspeakers

The many types and designs of domestic loudspeakers are familiar to hi-fi devotees and while the front-end antenna can mar the results, certainly the back end loudspeaker installation can significantly colour the results as heard by the motorist and passengers. The internal size and acoustics of the car will assuredly affect the sound quality, but the type of loudspeakers' how and where fitted are equally important.

Presuming that you do not wish to, or cannot for physical reasons, mount a conventional miniature sealed enclosure type loudspeaker in your car, a high grade unit should be mounted on a chipboard or hardboard baffle as large as is practicable ensuring that there is a good seal between the baffle board and its mounting. Metal panels do not make acceptable baffles due to resonance effects.

The position for siting loudspeakers must vary from car to car with the back shelf as probably the most common and the space beneath the dashboard extensively used. Some of the better 'pod' models are sealed units except for a small hole in the back linked to the inside of the cabinet by tubes or chambers. These can be placed on the back shelf as an alternative to flush mounting on a baffle. Other locations include the rear side panels, front door panels (combined tweeter mid-range units are available for this site) and when possible this is good for stereo sound, but not too low down or on the front kick panels for 'pod' units; flush mounted units just forward of the front panel.

For surround sound when back seat passengers are desirous of hearing clearly, several loudspeakers may be fitted. The overall impedance must match the tape player output and preferably a fader control of the correct type should be fitted to balance the levels of the front and rear speakers. If more than one loudspeaker unit is installed in the vehicle, they must also be wired in phase. Impedances that have been employed over the years range from 3 to 16Ω, but 4 and 8Ω are most common now.

Instead of the old faithful moving-coil type unit in some form or other, the Poly-Planar range uses an expanded polystyrene diaphragm which enables the familiar cone shape to be eliminated with a rigid flat panel acting as the transducer, not more than one inch in depth.

Traffic information services

A properly organised Traffic Information Service (TIS) and not merely announcements on local radio stations, in conjunction with special car radio circuits, must be in the offing to cope with Britain's growing traffic problem.

On the Continent, the most notable system is the German ARI (Autofahrer Rundfunk Information) which was devised by Bosch (Blaupunkt) some 10 years ago and has been used in Germany and Austria for a few years. The system has also been adopted by Denmark, and is being investigated by Yugoslav authorities and has been recommended by the Swiss police to its government.

If adopted here, the German ARI system would operate on the network of local VHF/FM stations (on BBC and IBA) rather than the main BBC VHF/FM services. The traffic information would form part of the normal programming, so it would not be possible to listen to national radio and receive ARI.

The full ARI system requires stereo operation of the transmitter as a 57kHz sub-carrier is radiated and modulated by low frequencies, typically 23·75, 28·27Hz etc., derived from the pilot carrier of 19kHz by sub-division.

These tones identify the ARI transmitter for the zone in which the car is travelling. A zone code letter is set up on a dial on the VHF radio (Fig. 17.10) and a panel LED lights up when the nearby ARI

Fig. 17.10 Blaupunkt Köln FM/AM stereo self-seek radio with ARI traffic information service selector under dash-board providing channel selection for different areas of West Germany

station is tuned in. In its basic format, the LED will illuminate on any ARI station received regardless of the zone location. The ARI decoder can mute or drop the normal radio programme level or that of a cassette unit in a car.

In West Germany figures quoted indicate 65% of cars fitted with VHF receivers, so it is not surprising that ARI was designed for that band and Blaupunkt's statistics claim that nearly 3,000,000 car radios are equipped with ARI. In Germany all the German manufactured cars factory-equipped with car radios also have integrated ARI decoders. Incidentally, a modified form of ARI can be used on the MW and LW bands. Bosch claim that a suitable ARI encoder and associated equipment could be fitted to the British existing VHF local station network at a cost of around £650 per installation and with minimal delay.

Carfax

In the UK towards the end of 1978 the BBC was experimentally testing its own Carfax system in the London area. The Carfax network would consist of remote controlled low power transmitters (50 to 100W) scattered around the country at existing BBC sites and some 40 to 70km apart, depending upon the area, all working on one common frequency. The frequency in use for these tests has been 526·5kHz or 1·5kHz just inside the bottom edge of the MW band. The use of a single discrete frequency is an essential feature of Carfax, originally proposed around 1973. The most important factor with this British system is that the traffic information is received independently of any other radio channel.

To receive Carfax transmissions users would have the option of buying a module (a small adaptor consisting of the front end of a MW receiver) for the Carfax frequency or getting a separate Carfax receiver with loudspeaker, or a new car radio with Carfax fitted. Carfax will discriminate against unwanted traffic information and the receiver would tune-in only to the nearest Carfax station broadcasting news about traffic conditions in the locality in which the motorist is driving at the time.

A switch selector on the radio would enable the motorist to arrange for the Carfax transmissions to automatically temporarily interrupt his normal radio or cassette tape listening, to receive Carfax transmissions only, or to use the car audio system with no Carfax reception at all. In practice only the strongest signal would switch the Carfax receiver and this would, in fact, be the required local transmitter.

Carfax has been costed at some £3,000,000 for a complete network, and most authorities agree it is the ideal traffic information service, but it must be on a discrete frequency and at present, the BBC's parlous finances would not allow them to sponsor this system, so presumably the initial costs would have to be borne by the Government or the motor/electronics industries? It will be necessary to bear in mind the international aspects of the system finally chosen, so as to make it workable in the UK and on the Continent. We understand that a combined Carfax/ARI decoder is technically feasible.

So although the ARI system would cost perhaps only £50,000 to install, traffic information would be an integral part of programming and thus heard by all listeners, not just by motorists, and any increase on the present level of traffic information transmitted during rush hours would become obtrusive. Coverage would also create difficulties since local radio does not yet provide countrywide service, and nor is it likely to. Again, where there is a choice, most motorists tend to listen to medium wave services which usually give better reception than VHF, and so many broadcasts would be missed. Finally, it would not be possible to listen to national radio such as Radio 1, and receive traffic information.

On the other hand, although Carfax would be expensive to set-up, since it is a separate dedicated radio channel, an even better motoring information service could be provided in heavily congested areas such as London where information is by necessity limited so as not to annoy non-motorists. Motorists would also be able to select any radio station and still receive Carfax without any special consideration as to the tuned channel.

Finally, if you are a dedicated 'in car' entertainment enthusiast with perhaps an interest in communications generally, you may wish to own a CB-Citizens' Band radio (for person to person communication over a limited area). Although not yet legal in the UK, quite a few transceivers from the USA and Japan are already in use in cars here. Again, if you are a 'ham' transmitter (amateur radio) you may wish to obtain a Sound Mobile licence. But that is another story (see Chapter 15).

Glossary of Terms and Abbreviations

Aerial Device usually mounted on the roof, for receiving (or transmitting) television or radio signals.

AM Amplitude modulation; a technique for impressing signals onto a carrier for transmission or recording.

ARI German traffic information system.

Bandwidth The frequency spectrum required to transmit or record television or audio signals.

Baud Effectively bits per second.

BBC British Broadcasting Corporation.

Beta Video cassette system developed by Sony.

Betamax Old name for video cassette system developed by Sony.

Black level The darkest part of a television picture representing no information.

bn billion.

BNC Type of video connector using half twist to lock.

Camera Device to produce television (or film) pictures.

Carfax BBC development for traffic information service.

Carrier Transmitted or recorded signal upon which other signals are often modulated or impressed.

Chromium dioxide Type of magnetic tape coating.

Coax plug/socket Type of aerial connector used on television sets.

CRT Cathode ray tube; display tube for television pictures.

CB Citizens band radio; loosely licensed radio.

Ceefax BBC name for teletext.

Chrominance Colour information in a television signal.

DIN Type of audio connector with multipins; based on West German standard.

Distortion When signals become unintentionally altered during recording or transmission, the result is often distortion.

Dolby Proprietary noise reduction system used on audio tape and cassettes.

Doppler shift Technique where frequencies change when interfered with by movement.

Drop out Expression for faulty magnetic tape coating that prevents normal signals being recorded.

Dubbing Technique of completing the sound track and effects on video tapes and film.

Edit Assembly of recorded or filmed sequences into desired order.

Eidophor Type of television projector producing large TV pictures.

Equalisation Altering the characteristics of a video or audio signal.

Firmware Computer program resident in a microcomputer.

FM Frequency modulation; technique for impressing signals onto a carrier for transmission or recording.

Field A part of a television frame, two fields to each frame.

Field strength Signal level at receiving aerial.

Format A standard specification, e.g. for video tape or cassette recording system.

Frame Each second 25 television frames are transmitted, each being a complete TV picture.

Framing control Control on video recorders that allows the machine to accommodate tapes recorded on other machines.

Ghosting Shadows or weak images on TV picture caused by reflections in air from buildings, hills, etc. or in cables.

GNP Gross national product; the total earnings of a country.

Guides Devices that identify the tape path in recorders.

Hardware Occasional term for equipment, particularly television.

Head drum Mechanism containing rotating heads in video recorders.

Helical Video tape recording system where tracks are recorded diagonally across the tape.

HF High frequencies.

Hz Hertz; frequency unit.

ICE In-Car entertainment.

IF Intermediate frequency used in television and radio receivers.

Impedance Measurement of matching on outputs and inputs of equipment.

Infra-red Light waves beyond the visible spectrum that are invisible to the naked eye but visible to detectors.

ITN Independent Television News.

ITV Independent Television (in Britain).

Jack Type of audio connector with single spigot.

Joystick Bidirectional control used in games allowing both vertical and horizontal movement on bat.

k thousand.

Laser Device producing sharply defined and intense beam of coherent light; used in some video disc players.

LCD Liquid crystal display; often used in watches with light absorbtion characteristics rather than light emitting.

LED Light emitting diode; type of solid state semiconductor light.

LF Low frequencies.

Line In British television, 625 lines make-up one television picture or frame.

Longitudinal Along the length of the tape.

Lumens Measure of light output from lamps and projectors.

Luminance Brightness (but not colour) information in a television signal.

LWT London Weekend Television.

M Million.

Market size Number of units that might be sold during a period; usually one year.

Microprocessor Semiconductor device containing thousands of transistors and being the principle component of a microcomputer.

Modified NTSC NTSC colour pictures replayed on PAL standard equipment giving a different colour standard; modified NTSC.

Modulator Device that impresses audio or video signals onto a carrier, which is often a television signal.

Monitor Type of television that accepts direct video and audio signals either instead of, or additionally to, an aerial input.

Monochrome Black and white; usually of television.

Music centre Unit which generally includes record deck, amplifier, radio and cassette deck mounted together.

Notch filter A device that removes certain frequencies often causing interference, from another signal.

NTSC Colour system used in Japan and USA.

OEM Original equipment manufacturer.

Oracle ITV name for teletext.

PAL Colour system used in Britain and most of Europe.

PCM Pulse code modulation; a digital technique for the transmission of sound.

Penetration   Total number of units in use.

PIR   Passive infra-red detector.

Phono   Type of audio connector with single spigot.

PO   British Post Office; previously GPO.

Polarisation   The direction in which a transmitting aerial (and thus the broadcast signal) is oriented.

Ports   In viewdata; the telephone lines terminate on ports which communicate with the computer.

PPM   Peak programme meter, type of level meter that reads peaks.

Prestel   Post Office trade name for viewdata.

Program   Instructions to a computer or microprocessor.

Programme   Pre-recorded television or film programming.

Programmable   (TV game) A game that offers a wide variety of options by plugging in different cartridges.

PSU   Power supply unit.

PVC   Polyvinylchloride (plastic).

Quadruplex   Professional video tape format using 2-inch wide tape.

RAM   Random access memory; semiconductor memory.

RIAA   Specialised equalisation used for record pick-ups.

ROM   Read only memory; semiconductor memory.

RPM   Revolutions per minute.

SECAM   Colour system used in France, USSR and Middle East.

Selectavision   Video disc format developed by RCA.

Semi-programmable   (TV game) A game that offers a restricted variety of options by plugging in different cartridges.

Software   Occasional term for television or computers program(me)s.

Splice   The joining of two ends of tape.

SVR   Super video recorder; video cassette format developed by Grundig.

Sync   (Television) Synchronising pulses are used to provide a stable reference to lock up TV pictures on a receiver or video recorder.

TeD   See Teldec.

Teldec   Video disc format developed by Telefunken and Decca.

Teletext   System of information transmission using normal television transmitters.

Telesoftware   Transmission of computer software (programs) by either teletext or viewdata.

TIS   Traffic information service.

Track Information is recorded on magnetic tape as tracks, of which there might be several on one tape.

Transistor Electronic device used for amplifying or switching signals.

Transmitter Device that allows television or sound signals to be broadcast.

Transposer Type of television replay transmitter.

Trinitron Television tube manufactured by Sony with colour stripes and in-line electron guns.

Tuner Device used to receive either television or radio transmissions.

UHF Ultra high frequencies; often used to indicate television bands 4 and 5 (colour).

UHF (F&E) Type of video connector using multiple twisting to lock.

U-Matic Industrial and commercial video cassette format developed by Sony.

Valve Old amplification device comprising electrodes in glass envelope.

VCR Trade name of Philips for video cassette recorder.

VCR-LP Long play video cassette recorder; trade name of Philips.

VDU Visual display unit; computer terminal.

VHF Very high frequency; often used to indicate television band 1 and 3, and radio band 2.

VHS Video home system; video cassette system developed by JVC.

Video cassette Plastic housing containing magnetic tape for recording television pictures.

Video cassette recorder Machine which records television programmes onto video cassettes.

Video disc Record on which television programmes are recorded.

Videogram Pre-recorded television programme made available on video disc or video cassette.

Vidicon Tube Type of camera tube normally only used for monochrome cameras.

Viewdata System of information transmission using telephone lines.

Viewfinder Device allowing camera operator to view camera image.

Visc Video disc developed by Matsushita (National Panasonic).

VLP   Video long playing disc; video disc developed by Philips.

VU   Volume units; type of level meter.

Zoom lens   Special type of lens whose focal length and thus length may be varied.

Index